Principles of Irrigation Engineering

Principles of Irrigation Engineering

Aaron Hall

New York

Published by Syrawood Publishing House,
750 Third Avenue, 9th Floor,
New York, NY 10017, USA
www.syrawoodpublishinghouse.com

Principles of Irrigation Engineering
Aaron Hall

International Standard Book Number: 978-1-64740-356-0 (Hardback)

Cataloging-in-Publication Data

Principles of irrigation engineering / Aaron Hall.
p. cm.
Includes bibliographical references and index.
ISBN 978-1-64740-356-0
1. Irrigation engineering. 2. Hydraulic engineering. 3. Agricultural engineering. I. Hall, Aaron.
TC805 .I77 2023
627.52--dc23

TABLE OF CONTENTS

PREFACE

Irrigation engineering is a branch of civil engineering that is involved in controlling and harnessing the various natural sources of water. This branch investigates various aspects of agriculture and irrigation in detail to determine the future prospects of irrigation. Irrigation engineering analyzes the efficiency of different irrigation systems to monitor their benefits and drawbacks. The main responsibility of irrigation engineering is to design and plan cost-effective and efficient irrigation systems. There are various advantages as well as disadvantages of developing irrigation systems but the benefits are far more than its disadvantages. One of the primary responsibilities of irrigation engineers deals with the problems that may arise in the watershed or the agricultural fields. In addition, irrigation engineers also deal with aspects such as the study of problems releated to water, soil, crop relationship, and the design and structure of dams, canals, and other hydraulic and irrigation structures. This book is compiled in such a manner, that it will provide in-depth knowledge about the theory and practice of irrigation engineering. With state-of-the-art inputs by acclaimed researchers in this field, it targets scholars and professionals.

After months of intensive research and writing, this book is the end result of all who devoted their time and efforts in the initiation and progress of this book. It will surely be a source of reference in enhancing the required knowledge of the new developments in the area. During the course of developing this book, certain measures such as accuracy, authenticity and research focused analytical studies were given preference in order to produce a comprehensive book in the area of study.

This book would not have been possible without the efforts of the authors and the publisher. I extend my sincere thanks to them. Secondly, I express my gratitude to my family and well-wishers. And most importantly, I thank my students for constantly expressing their willingness and curiosity in enhancing their knowledge in the field, which encourages me to take up further research projects for the advancement of the area.

Aaron Hall

1

Basic Concepts of Irrigation

1.1 Irrigation: Necessity, Advantages and Disadvantages

Objectives of Irrigation

Irrigation is facing new challenges that require refined management and innovative design. Formerly, emphasis was centered on project design. However, the current issues involve limited water supplies with several competing users, the threat of water quality degradation by excess irrigation and narrow economic margins. Meeting these challenges requires improved forecasting of irrigation water requirements.

Irrigation water requirements can be defined as the quantity or depth of irrigation water in addition to precipitation required to produce the desired crop yield and quality and to maintain an acceptable salinity balance in the root zone.

This quantity of water must be determined for uses such as irrigation scheduling for a particular field and seasonal water needs for planning, management and development of irrigation projects. The amount and time required for precipitation strongly influence irrigation water requirements.

In arid areas, annual precipitation is generally less than 10 inches and irrigation is necessary to successfully grow farm crops.

In semiarid areas, crops can be grown without irrigation but they are subject to droughts that reduce crop yields and can result in crop failure in extreme drought conditions.

Sub humid areas, which receive from 20 to 30 inches of annual precipitation are typically characterized by short dry periods.

Depending on the amount of water available, storage capacity of soils and crop rooting depth, irrigation may be required for short periods during the growing season in these areas. In humid areas, those receiving more than 30 inches of annual precipitation, it normally exceeds transpiration throughout most of the year.

However, yield decreases due to the occurrence of drought periods which decreases and impair quality especially for crops grown on shallow, sandy soils or that have a shallow root system. Irrigation is not needed to produce a crop but may be required to protect against an occasional crop failure and to maintain product quality.

Necessity of Irrigation in India

There is a great necessity of irrigation in Indian agriculture. India has a great diversity and variation in climate and weather conditions. These conditions range from extreme heat to extreme cold and from extreme dryness to heavy rainfall. Due to some reasons irrigation is needed in Indian agriculture:

- Irregularity in distribution of rainfall throughout the year.
- Uncertainty of Monsoon rainfall both in time and place.
- Drought is an annual event in some areas.
- Excessive rainfall causing flood.
- India is a land of Rabi Crops. But there is not rainfall in winter months.
- Some soil need more water.
- Introduction of H.Y.V seeds and multiple cropping need water throughout the year.

1. Variety of Climate: The climatic and weather conditions of India Varies in wide range. There is extreme temperature at some places, while the climate remains extremely cold at other places. While there is excessive rainfall at some places, other places experience extreme drought. So, irrigation is needed in India.

2. Irregular and uncertain monsoon: India is country of monsoon. But monsoon is irregular and erratic in nature. Sometimes it comes easily and brings heavy rainfall, but sometimes it gets delayed and brings inadequate rainfall. Further, there is irregularity in the distribution of rainfall all through the year. The irrigation system helps the farmers to have less dependency on rain-water. During the months when there is scarcity in rainfall, the crops are supplied water through irrigation systems.

3. Agriculture based economy: Indian economy is predominantly dependent on agriculture. A large portion of Indian population depends on agriculture. Without irrigation agriculture is a difficult task in dry areas or during the months of inadequate rainfall. Naturally, for the agricultural activities across the wast diverse regions, there is a need for proper irrigation system.

4. Winter crop: India is a vast country with rich luminous plain lands suitable for Rabi as well Kharif crops. But there is no rainfall during winter months in many places. Without irrigation cultivation of Rabi crops along with other crops is not possible. Rabi crops are grown during the long dry spell of winter season. It is mainly possible due to advanced irrigation facilities.

5. High breed seeds: At present because of hybrid seeds, crops can be produced at any seasons. But the production of crops predominantly depends on water. The introduction

of advanced irrigation system have enabled the farmers to produce crops even during dry season.

6. Soil Character: In many places, the soils have less water-retaining capacities.

7. Jute and Paddy: Irrigation is needed for growing some "water observing plants like jute and paddy.

The types of Irrigation mainly practiced in India are:

- Tanks:
  - Sichhni.
  - Donga.
- Well:
  - Dug Well.
  - Tube Well:
    - Shallow.
    - Deep.
- Canal:
  - Perennial.
  - Non-Perennial.

Advantages of Irrigation

- During the period of low rainfall or drought yield of crops may increase or remains same due to irrigation system.
- The food production of a country will be improved by ensuring the growth of crops. This helps a country to prevent famine situation.
- Securing increased agricultural production and thus improving the nutrition of the population.
- Irrigation helps to improve the cultivation of cash crops such as vegetables, fruits, tobaccos, sugar cane.
- In some river valley projects multipurpose reservoirs are formed by constructing high dams. At these river valleys, hydroelectric power might be generated.

- Retention of water in reservoirs and the use of possible multipurpose.
- Irrigation canal might be the source of water supply for domestic and industrial purposes.
- The reservoirs and canals will be utilized for the development of the fisher project.
- Culturing the area increasing the social and cultural level of the population.
- Recreation facilities in irrigation canals and reservoirs.
- Increases employment by providing jobs to people.
- Improvement of the micro climate. Possibility provided for waste water use and disposal.
- Improvement of water regime of the irrigated soils.

Disadvantages of Irrigation

- Danger of water logging and salinization of soils.
- It could change properties of water in reservoirs due to waste water use and disposal.
- Deforestation of area is to be done that is to be irrigated.
- Possible spread of diseases from certain types of surface irrigation.
- Danger of pollution of water resources by return run off from irrigation.
- New diseases caused by retention of waste water in large reservoirs.
- Due to excessive irrigation climate becomes damp and cold. Thus humidity increases which is not good for health.
- Careless irrigation might lead to retention of water and create places for breeding of mosquitoes.
- Excess of irrigation might result in raising the sub soil water table and lead to water logging of the area.

1.2 Techniques of Water Distribution in Firms

It is essential to understand the basic types of water supply systems and the physical arrangement of system components prior to discussing either design or evaluation concepts associated with these components.

There are two basic types of water supply systems to create water pressure within the distribution system to supply water to the built areas of a community and to provide required water pressure to fire hydrants located on the water supply system. These two types of water systems that supply water under pressure for consumer consumption and fire protection are:

- Gravity feed systems.
- Pumping pressure systems.

Each of these systems must take water from a supply source, pass the water through a treatment plant and then transport the water into the distribution system. It needs to be recognized that community water systems can be divided into four basic classification according to the water source:

- High or low reservoirs that hold non-potable water for gravity feed.
- Pumping station systems that use ground water from streams, rivers, canals, man-made or natural lakes and other special provisions for impound water. In these types of systems, the raw water is pumped from the source point to the treatment plant and then either pumped directly into the distribution system or into storage to be used on demand by the community.
- Pumps at well sites that pump water to the treatment facility. Based on the difference in elevation between the treatment facility and the community to be served, the water may flow by gravity through the distribution systems or there may be the need for another pumping station.
- A combination of gravity flow and one or more pumping stations to transport the water from the source point to all of the water demand points on the distribution system.

1.2.1 Quality of Irrigation Water

A good irrigation water is the one which performs the functions without any side effects which retard the plant growth.

Irrigation water may be said to be unsatisfactory for its intended use if it contain:

- Chemicals toxic to plants or the persons using plant as food.
- Chemicals which react with the soil to produce unsatisfactory moisture characteristics.
- Bacteria injurious to persons or animals eating plants irrigated with the water.

Impurities in Irrigation Water

The quality of irrigation water depends upon various types of impurities present in water, the following being the prominent ones:

- Concentration of sediments in water.
- Total concentration of soluble salts (known as TDS).
- Proportion of sodium ions to other cations.
- Concentration of toxic elements such as boron concentration.
- Concentration of bicarbonate in relation to the concentration of calcium and magnesium.
- Bacterial concentration.

The effect of concentration of sediments transported by irrigation water depends upon the type of soil of irrigation fields. If the field contains sandy soil, its fertility is improved by the deposition of sediments carried by irrigation water. However, if the sediment has been derived from eroded areas, if may reduce the fertility and decrease the permeability of soil.

High concentration of salts in water may be harmful in the long run. Similarly, if water contains high concentration of sodium ions, the irrigation soils become plastic and sticky when wet and become prone to form clouds and they crust on drying. Similarly high percentage of toxic elements may be harmful to certain plants such as nuts, citrus fruits and deciduous plants.

High concentration of bicarbonate ions may result in precipitation of calcium and magnesium bicarbonates from the soil solution, increasing the relative proportion of sodium irons and causing sodium hazards. Bacterial contamination may not be directly harmful to the plants as such, but it may be injurious to persons or animals eating these plants.

1.3 Water Requirements for Crops

Water requirements for crops vary from place to place depending upon nature of crops and site. In some areas there is no need of irrigation because the conditions are fulfilled by natural resource i,e, rainfall.

It is the water required by the plants for its survival, growth, development and to produce economic parts. This requirement can be fulfilled either naturally through precipitation or artificially through irrigation.

Crop water requirement includes all losses like:

- Transpiration loss that occurs through leaves (T).
- Evaporation loss through soil surface in cropped area (E).
- Certain application losses includes conveyance loss, percolation loss, runoff loss, etc., (WL).
- Amount of weather consumed by plants (WP) for its metabolic activities that is estimated to be less than 1% of the total water absorption. The separation of these three components are not so easy. ET loss is taken as crop water use or crop water consumptive use.
- The water needed for special purposes (WSP) like puddling operation, ploughing operation, land preparation, leaching requirement, weeding, dissolving fertilizer and chemical, etc.

Water requirement is symbolically represented as:

$$WR=T+E+WP+WL+WSP.$$

The estimations of the water requirement of crop are one of the basic needs for crop planning on the farm and for planning of any irrigation project.

Water requirement might be defined as the quantity of water required by a crop or diversified pattern of crop in a given period of time for its normal growth under field conditions at a place.

Water requirements includes losses due to ET or CU and losses during the application of irrigation water and also the quantity of water required for special purposes such as operations like land preparation, transplanting, leaching etc.

It can be formulated as WR = ET or Cu + application loss + water for special needs. It can be stated based on "Demand" and "supply source" as follows,

$$WR=IR+ER+S$$

Where,

IR-Irrigation requirement.

ER-Effective rainfall.

S - Contribution from ground water table.

The idea about crop water requirement is essential for farm planning with respect to total quantity of water required and its efficient use for various cropping schemes of the farm or project area. Crop water requirement is also to decide the stream size and design the canal capacity.

Combined loss of evaporation and transpiration from a cropped field is called as evapotranspiration. Otherwise known as consumptive use and denoted as ET and this is a part of water requirement,

$$CU = E + T + WP$$

Therefore,

$$WR = CU + WL + WSP$$

Crop water requirement can also be defined as water required for meeting evapotranspiration demand of the crop and special needs in case of wet land crop. It also includes other application losses both in the case of wet land and garden land crops. Also known as crop water demand.

The requirement of crop water varies from place to place, from crop to crop and depends on agro-ecological variation and crop characters.

The following features that mainly influence the crop water requirement are:

- Crop factors:
  - Growth stages.
  - Variety.
  - Plant population.
  - Duration.
  - Crop growing season.
- Soil factors:
  - Structure.
  - Depth.
  - Texture.
  - Soil chemical composition.
  - Topography.
- Climatic factors:
  - Wind velocity.
  - Relative humidity.

    - Sunshine hours.
    - Rainfall.
    - Temperature.
- Agronomic management factors:
    - Irrigation methods used.
    - Frequency of irrigation and its efficiency.
    - Tillage and other cultural operations like weeding, mulching etc./inter-cropping etc.

Based on all these factors, average crop water requirement for various crops have been worked out.

Crops and Crop Season

India is top producer country of many crops. The crops can be classified based on:

- Agricultural.
- Crop seasons.
- Irrigation requirement.

Principal Crops of India

The below table gives the principal crops of India, along with crop season and average depth of water required.

Crop	Sowing time	Harvesting time	Average Delta (cm)
(A) Kharif Crop			
Rice	June-July	Oct -Nov.	120
Makai(Maize)	June-July	Sept-Oct.	45
Bajra(Spiked Millet)	June-Aug	Sept-Oct.	30
Jowar (Great Millet)	June-July	Oct-Nov.	30
Pulses(Mung, Moth, Urd)	June-July		30
Ground Nut	May		45

(B) Rabi Crops			
Gram	Sept-Oct.	March-April	30
Wheat	Oct-Nov.	March-April	40
Barley	Oct-Nov.	March-April	45
Peas	Oct-Nov.	March-April	50
Mustard	Oct.	Feb- March	45
Tobacco	Feb-March	June	60
Potato	Oct.	Feb.	75
(C) Eight Months Crops			
Cotton	May-June	Dec-April	45
(D) Perennial Crops			
Sugar-cane	Feb-March	Dec - March	80

1. Classification Based on Agriculture

Agriculture consists of the following types of crops:

- Field crops: Such as wheat, rice, maize, barley, oats, great millet, spiked millet, gram, pulses etc.
- Commercial crops: Such as sugarcane, cotton, tobacco, hemp, sugar beat etc.
- Oil seed crops: Such as mustard, ground nut, sesame, linseed, caster etc.
- Horticulture crops: Consisting of various fruit crops, various vegetable crops and flower crops.
- Plantation crops: Such as tea, coffee, cocao, coconut, rubber etc.
- Forage crops: Such as fodder, grass etc.
- Miscellaneous crops: Such as medicinal crops, aromatic crops, sericulture crops, condiments and spices.

Major Crops

The crops are divided into two major groups:

- Food grains.
- Non-food grains.

Food Crops

Rice

Rice is India's staple food. Next to China, India is one of the most leading producer of rice in the world. Geographical conditions required are clayey soil and standing water during growth.

Temperature to be uniformly high (above 25°C) and rainfall should be between 100 to 200 cm.

It is a kharif crop. Rice cultivation is concentrated mainly in the Northern plains which have alluvial soils and adequate water supply. West Bengal is the leading producer of rice.

Rice is India's staple food. Next to China, India is one of the most leading producers of rice in the world. Geographical conditions required are clayey soil and standing water during growth.

Temperature to be uniformly high (above 25°C) and rainfall should be between 100 to 200 cm.

It is a kharif crop. Rice cultivation is concentrated mainly in the Northern plains which have alluvial soils and adequate water supply. West Bengal is the leading producer of rice.

Wheat

It is one of oldest crop introduced in India from Middle East. It is a rabi or winter crop and grown in winter season when the temperature is lesser than 20°C. Annual rainfall of 50 to 75 cm is adequate for the crop.

Loamy soils of Northern plains and black soils of Deccan are suited for wheat cultivation. Wheat is sown generally in October-November and harvested in March-April. India stands fourth in wheat production with about one-eighth of world output.

Millets

Millets are coarse grains which serve as food for a large number of people in India. They are kharif crops and grow in less rainy areas in the following order - Ragi (damp area), Jowar (moist area) and Bajra (dry area).

Millets require high temperature and less rainfall. They are the alternative to rice as rainfall decreases. Ragi is confined to areas such as Karnataka and Tamil Nadu, Jowar in Karnataka, Andhra Pradesh, Maharashtra and Bajra in drier parts of Maharashtra, Gujarat, Rajasthan and south west Uttar Pradesh. India stands first in production of millets.

Maize

Maize, being an American crop, is a relatively new entrant and is gaining popularity because of its high yields and its easy adaptability to different soils and climatic conditions. Rich in protein and requires moderate rainfall.

Pulses

India is the largest producer and also the largest consumer of pulses. They are rich in protein. It requires moderate rainfall and verifying temperature.

Pulses include gram, arhar, moong, peas, masoor and leguminous plants. They help in restoring the fertility of the soil.

Non-Food Crops

Oil Seeds

India is one of the leading producers of oil seeds in the world. Main source of edible oils. Oil seeds are used for preparing paints, varnishes, perfumes, medicines, soap etc.

Main oil seeds are groundnut (kharif crop in peninsular India), rapeseed and mustard (rabi crops in wheat belt). Additional oilseeds are seasamum (Orissa, Rajasthan, West Bengal, Tamil nadu, Maharashtra), Linseed (Madhya Pradesh, Uttar Pradesh and Maharashtra), Castor seed (Gujarat) and Cotton Seed (Maharashtra, Punjab).

India is the world's third largest producer of coconut, of which Kerala accounts for two-thirds of total production.

Oilseeds	1998-99	1999-2000	(Estimated) 2000-01
Groundnut	9.0	5.3	6.2
Rapeseed	5.7	6.0	4.3
Soyabean	7.1	6.8	5.2
Other 6 oilseeds	2.9	2.8	2.9
Total 9 oilseeds	24.7	20.9	18.6

Sugarcane

India is known as the original land of sugarcane. It is sown just before kharif season and harvested in winter.

Sugarcane requires about 100 cm of rain. New varieties of sugar such as khandsari and gur are produced from sugarcane.

Per Hectare	Productivity in India	
Crop / Item	Per hectare Production (in Kg)	
	1998-99	1999-2000
Total food grains	1571	1697
Cereals	1778	1919
Pulses	634	630
Rice (Kharif)	1619	1886
Rice (Rabi)	3073	2968
Rice (Total)	1747	1990
Wheat	2590	2755
Maize	1797	1785
Oilseeds (Kharif)	992	809
Oilseeds (Rabi)	868	935
Oilseeds (total)	944	856
Groundnut (Kharif)	1174	665
Groundnut (Rabi)	1370	1389
Rape seed	870	982
Sugarcane	71	71
Cotton	224	226
Jute	1875	1995
Mesta	990	1078

Fibres

Cotton

It is a kharif crop and is known as the king among fibres. India stands fourth in the total world's production. It requires warm climate and high temperature. There are three varieties: Long staple, Medium staple and Short staple.

India mostly produces medium and short staple. Gujarat and Maharashtra are some of the leading producers of cotton.

Jute

Jute was called the golden fibre of Indian sub-continent. India ranks second in the world's production of Jute in the world after Bangladesh.

Jute is a fibre plant and it requires high temperature and rainfall above 200 cm. Grows well in flood plains or in the well-drained fertile soil.

Agricultural Production

Crops	1997-98	1998-99	1999-2000	2000-01
Rice	82.5	86.0	89.5	86.7
Wheat Coarse	66.3	71.3	75.6	70.0
Cereals	30.4	31.2	30.5	29.9
Pulses	13.0	14.9	13.4	11.7
Total food-grains	192.3	203.5	208.9	198.3
Kharif food-grains	101.6	102.8	104.9	102.0
Rabi food-grains	90.7	100.7	104.0	96.3
Oil seeds	21.3	24.7	20.9	18.6
Sugarcane	279.5	288.7	299.2	300.6
Cotton	10.9	12.3	11.6	13.2
Jute/Mesta	11.0	9.8	10.5	9.9

Plantation Crops

Tea

India is the leading producer of tea. It requires a temperature of 24°C to 30°C.Rainfall on an average should be above 200cm. Soil should be deep fertile and well drained where water stagnation does not take place.

Production, Consumption and Export of Tea

Year	Production	Consumption	Export
1996	7800	618	161.7
1997	810.6	640	203.0
1998	870.4	645	210.3
1999	806.0	655	190.0
2000	784.0	NA	178.0

Undulating plains of Brahmaputra and Surma Valleys of Assam is known as the house of Indian tea. Tea is one of the major foreign exchange earners of India.

Coffee

It stands second as a popular beverage in the world as well as in India. Thrives in rich and well-drained soil and it grows best in tropical highlands.

If tea belongs to north-eastern part, coffee is confined to the south-western part. Nearly 50% of Indian's production is exported to various countries.

Rubber

Rubber plantations were first established in Kerala in 1902. Needs hot-wet climatic conditions. Most of the land under rubber belongs to small land holders.

Fruits

A large variety of fruits are produced in India. This includes Bananas, Pineapples, Jack-fruits, and Oranges that are grown in tropical region.

India is a leading producer in Cashew nut. Apples, Plums, Peaches, Almonds which are grown in Himachal Pradesh. Fruits are also exported. India ranks second in fruit production.

Tobacco

India is fourth largest producer and sixth larger exporter of tobacco. Crop needs freedom from frost. Tobacco needs fertile soils and heavy doses of fertilizers. Andhra pradesh and Gujarat are leading producers of tobacco.

2. Classification Based on Crop Seasons

Based on crop season, crops are classified as follows:

- Rabi crops or Winter Crops: These crops are sown in autumn (or October) and are harvested in spring (or March). Various crops that fall under this category are: gram, wheat, barely, peas, mustard, tobacco, linseed, potato etc.
- Kharif crops or monsoon crops: These crops are sown by the beginning of the southwest monsoon and are harvested in autumn. These consist of rice, maize, spiked millet, great millet, pulses, groundnut etc.
- Perennial crops: These are the crops that require water for irrigation throughout the year. Examples of perennial crops are: sugar cane, fruits, vegetables etc.
- Eight months crops: These crops, such as cotton, require irrigation water for 8 months.

Agricultural Seasons in India

i. The Kharif Season

- Crops are sown at the starting of south-west monsoon and harvested at the end of the south-west monsoon.
- Sowing Season: May to July.
- Harvesting Season: September to October.
- Important Crops: Jowar, Bajra, Rice, Maize, Cotton, Groundnut, Jute, Hemp, Tobacco etc.

ii. The Rabi Season

- During the growth period, the crops need cool climate and requires warm climate during the germination of seed and maturation.
- Sowing Season: October to December.
- Harvesting Season: February to April.
- Important Crops: Wheat, Barley, Mustard, Gram, Linseed, Masoor & Peas.

iii. The Zaid Season

These Crops are raised throughout the year through artificial irrigation:

- Zaid Rabi Crops:
    - Sowing Season: February to March.
    - Harvesting Season: April-May.
    - Important Crops: Watermelon, Toris, Cucumber & other vegetables.
- Zaid Kharif Crops:
    - Sowing Season: August to September.
    - Harvesting Season: December-January.
    - Important Crops: Rice, Jowar, Rapeseed, Cotton, Oil seeds.

3. Classification Based on Irrigation Requirements

Based on irrigation requirements, crops can be classified as:

- Dry crops.

- Wet crops.
- Garden crops.

Dry crops are the one which do not require water for irrigation where only rain water is sufficient for their growth. Wet crops are those which cannot grow without irrigation. Garden crops require irrigation throughout the year.

TRIFED

Tribal Co-operative Marketing Development Federation of India Ltd. (TRIFED), was established in August 1987 and functioning started in 1988. Aim was to save tribal from exploitation by private traders and to offer remunerative prices for their minor forest produce.

TRIFED is an agency for collecting, processing, storing and developing all oil seed products. Agriculture ministry gives aid to TRIFED for compensating loss incurred due to price fluctuations.

NAFED

The National Agricultural Cooperative Marketing Federation of India Ltd. (NAFED), was established in cooperative sector at national level for the marketing of agricultural products.

Duty

Duty represents the irrigating capacity of a unit water. It relates the area of a crop irrigated and the quantity of irrigation water required during crop growth.

For example: If 5 cumec of water is required for a crop sown in an area of 5000 hectares, The duty will be 5000/5=1000 hectares/cumec.

Delta

Delta is total depth of water required for a crop during the entire period the crop is in the field Denoted by Δ. The unit of delta is days.

Base Period

It is the whole period from irrigation water first issued for preparation of the ground for planting the crop to its last watering before harvesting.

Relation between Duty and Delta

$$\text{delta}(\Delta) = 8.64\frac{B}{D}$$

Where,

Δ = Delta in meter.

B = Base period in days.

D = Duty in hectares/cumec.

Factors Affecting Duty

The duty of water of canal system depends upon a variety of the factors. The principal factors are:

- Methods and system of irrigation.
- Mode of applying water to the crops.
- Method of cultivation.
- Time and frequency of tilling.
- Type of the crop.
- Base period of the crop.
- Climatic conditions of the area.
- Quality of water.
- Method of assessment of irrigation method.
- Canal conditions.
- Character of soil and sub-soil of the canal.
- Character of soil and sub-soil of the irrigation fields.

1. Methods and Systems of Irrigation: In the perennial irrigation system, soil is continuously kept moist and hence water required for initial saturation is less. Also, due to the shallow depth of the water table, deep percolation losses are less. In the inundation irrigation, there is wasteful use of water. Hence, the perennial irrigation system has more duty than the inundation irrigation.

The flow irrigation system has lower duty due to the conveyance losses in the network of the canals, while the lift irrigation system has higher duty because the commanded area of each well is very near to it. Tank irrigation gives high duty due to rigid control.

2. Mode of Applying Water: The flood irrigation system has lesser duty than the furrow system. Sub-irrigation system gives still higher duty. The ring basin irrigation and uncontrolled flooding give less duty.

3. Method of Cultivation: If the land is properly ploughed and made quite loose before irrigating, the soil will have high water retention capacity in its unsaturated zone. Thus, the number of waterings can, be reduced, increasing the duty. The old and conventional methods of cultivation gives less duty in comparison to the modem methods.

4. Time and frequency of Tilling: Frequency of cultivation reduces the loss of moisture through weeds. Soil structure affects the plant growth to a very great extent. A good structure (i.e. the good arrangement of soil particles in relation to one another) is called good tilth of soils. When the soil is in good tilth, evaporation losses from the surface of soil is less, soil becomes properly aerated and hence the yield of crop is also better.

5. Type of the crop: The duty varies from crop to crop.

6. Base Period of the Crop: If the base period of the crop is more, the amount of water required will be high, hence duty will be low and vice-versa.

7. Climatic Conditions of the Area: The climatic conditions which affect the duty are: (i) temperature, (ii) wind, (iii) humidity and (iv) rainfall. Due to high temperature and wind, evaporation losses will be more and duty will be less. A humid atmosphere reduces the losses. Rainfall during the crop-period will reduce the irrigation-water requirement and the duty will thus be higher. In this context there are two popular terms:

- Duty inclusive of rainfall, i.e. duty be considered by taking rainfall into account.
- Duty exclusive of rainfall, that is the duty has been considered by not considering the rainfall in the area.

8. Quality of water: If the harmful salt-content and alkali content of the water is more, water will have to be applied liberally so that the salts are leached off. This will reduce the duty. More fertilizing matter in water will cause less consumption of water and increase duty.

9. Method of Assessment of water: Volumetric method of assessment always leads to a higher duty. This is because the farmer will use water economically. If the method of assessment is based on the area under cultivation, the farmer will have a tendency to use more water and the duty will be low.

10. Canal Conditions: In an earthen canal, the seepage and percolation losses will be high, resulting in the low duty. If, however, the canal is lined, the losses will be less and duty will be high. If the high canal is so aligned that the irrigated areas are concentrated along it, the duty will be higher. The dispersion of the irrigated areas with respect to the canal tends to reduce the duty.

11. Character of Soil and Sub-soil of the Canal: If the canal is unlined and if it flows through coarse grained, permeable soils, the seepage and percolation losses will be high. If the canal flows through fine grained soil, such losses will be less and hence the duty will be higher.

12. Character of Soil and Sub-soil of the Irrigation Field: If the soil and sub-soil of the field is coarse-grained, percolation losses will be high. However, if there is hard pan at depth 1 to 2 meter below surface, the percolation loss reduces. The duty is also affected by the topography of the land. If the field is not level, the lower portions get more water, while higher portions may remain drier. In order to supply water to the higher places, more water will be used and the duty will be reduced.

1.4 Consumptive Water use and Irrigation Requirements

Consumptive Water Use

Consumptive water use is water removed from available supplies without return to a water resource system. Evaporation from the surface of the earth into clouds of water in the air which then falls to the ground as “rain” is excluded from this model.

Crop consumptive water use is the amount of water transpired during plant growth plus what evaporates from the soil surface and foliage within the crop area. The portion of water consumed in crop production depends on many factors especially the irrigation technology.

It is the quantity of water required by the vegetation growth of a given area. The vegetation growth includes evapotranspiration and also for the building of plant tissues plus evaporation from soils and intercepted precipitation. It is expressed in terms of depth of water.

Consumptive use varying factors includes temperature, humidity, wind, speed, topography, sunlight hours, method of irrigation, moisture availability.

Mathematically,

$$\text{Consumptive Use} = \text{Evapotranspiration} = \text{Evaporation} + \text{transpiration}$$

Factors Affecting the Consumptive Use of Water

The Consumptive use of water changes with:

- Evaporation which depends on humidity.
- Mean temperature of the month.
- Growing season of crops and the cropping pattern followed.
- Monthly precipitation over an area.
- Wind velocity in locality.
- Soil and topography.

- Irrigation practiced and method of irrigation.
- Sunlight hours.

Types of Consumptive Water Use:

- Optimum Consumptive Use.
- Potential Consumptive Use.
- Seasonal Consumptive Use.

1. Optimum Consumptive Use: It is the consumptive use that produces a maximum crop yield.

2. Potential Consumptive Use: If sufficient moisture is always available to completely meet the requirements of vegetation fully covering the entire area then resulting evapotranspiration is known as Potential Consumptive Use.

3. Seasonal Consumptive Use: During entire growing season, the total amount of water consumed in the evapotranspiration by a cropped area accounts to seasonal consumptive use.

Irrigation Requirements

The major objectives are:

- To supply water partially or totally for crop need.
- To cool both the soil and the plant.
- To leach excess salts.
- To improve groundwater storage.
- To facilitate continuous cropping.
- To enhance fertilizer application-Fertigation.

As the population is increasing day by day and demand of more food irrigation has become necessary and if the area is in arid zone and do not have evenly distributed rainfall then this large area can be brought under cultivation because of irrigation by means of artificial application of water.

Irrigation is needed for the following purposes:

- Irrigation is required for normal growth and yield of the plants and crops.
- For metabolic processes of the plant.

- To reduce the soil temperature.
- For easy germination of seed from the soil.
- Irrigation water acts as a medium for the transport of nutrients and photosynthesis in the plant system.
- To provide insurance against short duration drought.
- To wash out and dilute the salts in the soil.
- To reduce the hazard of soil piping.
- To soften tillage pans and clods.
- Factors governing necessity of irrigation.
- Washing or diluting salts in the soil.
- Delaying bud formation by evaporative cooling.
- Promoting the function of some microorganisms.

The following are the factors for determining the need for irrigation:

- Insufficient rainfall.
- Un-even distribution of rainfall.
- Improvement of perennial crops.
- Improving the number of crops during the year.
- Development of agriculture in desert area.
- Controlled water supply.
- Commercial crops with additional water.

Estimation of Consumptive use of Water by Climatic Approaches

To measure or estimate the consumptive use the following methods are used:

- Direct Methods/Field Methods.
- Empirical Methods.
- Pan evaporation method.

Direct Methods

In this method field observations are made and the physical model is used for this purpose. This includes:

- Vapour Transfer Method/Soil Moisture Studies.
- Field Plot Method.
- Tanks and Lysimeter.
- Integration Method/Summation Method.
- Irrigation Method.
- Inflow Outflow Method.

1. Vapour Transfer Method

It is used for estimation of water consumptive use. Soil moisture measurements are taken before and after each irrigation. The quantity of water extracted per day from soil is computed for each period.

2. Field Plot Method

It Select a representative plot of area and the accuracy depending upon the representativeness of plot (cropping intensity, exposure etc.). It replicates the conditions of an actual sample field (field plot). Less seepage should be there.

$$\text{Inflow} + \text{Rain} + \text{Outflow} = \text{Evapotranspiration}$$

The drawback in this method is that lateral movement of water takes place although more representative to field condition. Some correction has to be applied for deep percolation as it cannot be ascertained in the field.

3. Tanks and Lysimeter

A watertight tank of cylindrical shape having diameter 2m and depth about 3m is placed vertically on the ground to measure consumptive use of water. Tank is filled with sample of soil. Bottom of the tank consists of a sand layer and a pan for collecting the surplus water.

Plants grown in the Lysimeter must be the same as in the surrounding field. Consumptive use of water is estimated by measuring the amount of water needed for satisfactory growth of plants within the tanks.

Consumptive use of water is given by,

$$C_u = W_a - W_d$$

Where,

C_u = Consumptive use of water.

W_a = Water Applied.

W_d = Water drained off.

Lysimeter studies are time consuming and expensive. Methods 1 and 2 are the more reliable methods as compare to this method.

4. Integration Method

In this method, it is necessary to know the division of total area, i.e. under irrigated crops, natural native vegetation area, water surface area and bare land area. Here annual consumptive use for the whole area is found in terms of volume. Expressed in Acre feet or Hectare meter.

Mathematically,

Total Evapotranspiration = Total consumptive use.

Total Area Annual Consumptive Use = Total Evapotranspiration = A + B + C + D

Where,

A = Unit consumptive use for each crop * its area.

B = Unit consumptive use of native vegetation * its area.

C = Water surface evaporation * its area.

D = Bare land evaporation * its area.

5. Irrigation Method

In this method, unit consumption is multiplied by some factor. Multiplication values depend upon the type of crops in certain area. This method requires an Engineer judgment as these factors requires Engineer's investigation of certain area.

6. Inflow Outflow Method

For large areas annual consumptive use is found in this method.

If U is the valley consumptive use its value is given by,

$$U = (I + P) + (G_s - G_e) - R$$

Where,

U = Valley consumptive use (in acre feet or hectare meter).

I = Total inflow during a year.

P = Yearly precipitation on valley floor.

G_s = Ground Storage at the beginning of the year.

G_e = Ground Storage at the end of the year.

R = Yearly Outflow.

All the above volumes are measured in acre-feet or hectare-meter.

Empirical Methods

Empirical equations are given for the water requirement estimation. These are,

Lowry Johnson Method

The equation for this method is,

$U = 0.0015\ H + 0.9$ (Over specified)

U = Consumptive Use.

H = Accumulated degree days during the growing season calculated from maximum temperature above 32 °F.

Penman Equation

According to this method,

$U = E_T = AH + 0.27\ E_a A - 0.27.$

E_T = Consumptive use or Evapotranspiration in mm.

E_a = Evaporation (mm/day).

H = Daily head budget at surface (mm/day).

H = Function of sunshine hours, radiation, wind speed, vapour pressure and other climatic factors.

A = Saturated vapour pressure curve slope of air at absolute temperature in °F.

Hargreave's Method

It is a very simple method.

According to this method,

$Cu = K * E_p$

Where,

Cu = Consumptive Use coefficient (varies from crop to crop).

E_p = Evapotranspiration.

K = Coefficient.

Pan Evaporation Method

It is a measurement that combines or integrates the effects of some climate elements like temperature, humidity, rainfall, drought dispersion, solar radiation and wind.

Evaporation takes place to a greater extent on hot, windy, dry, sunny days and is greatly reduced when clouds block the sun and when air is cool, calm and humid. Enable farmers and ranchers to how much water their crops may need.

1.5 Irrigation Efficiencies and Soil Moisture-Irrigation Relationship

Irrigation efficiency is the ratio between the water stored in the soil depth inhabited with active plant roots to the water applied by the irrigation system. Thus water applied by the irrigation system are not being made available to be taken up by plant roots is wasted and reduces irrigation efficiency.

Major causes for reduced irrigation efficiency are drainage of excess irrigation water to soil layers deeper than the depth of active roots. Leakage of irrigation water to deep soil layers could result in pollution of the water table.

Irrigation efficiency of 100 percent are practically nonexistent even in the most modern irrigation systems. Conservative estimates suggest that even under optimal management practices the average irrigation efficiency is estimated to be 70 percent.

Average water loss under sprinkler and drip irrigation is 30 percent but could drop to values of over 50 percent under furrow and flood irrigation.

Water losses of irrigation water under urban and landscape irrigation could easily reach 50 percent of the applied water. In Israel, a yearly saving of 300 to 400 million metric volumes of irrigation water could be saved as a result of using technologies capable of increasing substantially irrigation efficiencies.

Soil Moisture-Irrigation Relationship

Any given volume V of soil consists of:

- Volume of solids V_s.

- Volume of liquids V_w.
- Volume of gas V_a.

Obviously, the volume of voids $V_v = V_w + V_a$. For a fully saturated soil sample, $V_a = 0$ and $V_v = V_w$. Likewise, for a completely dry specimen, $V_w = 0$ and $V_v = V_a$. The weight of air is considered zero compared to the weights of water and soil grains. The void ratio e, the porosity n, the volumetric moisture content w and the saturation S are defined as,

$$e = \frac{V_u}{V_s}, n = \frac{V_u}{V}, w \frac{V_w}{V}, S = \frac{V_w}{V_u}$$

Therefore,

$$\omega = Sn \qquad ...(1)$$

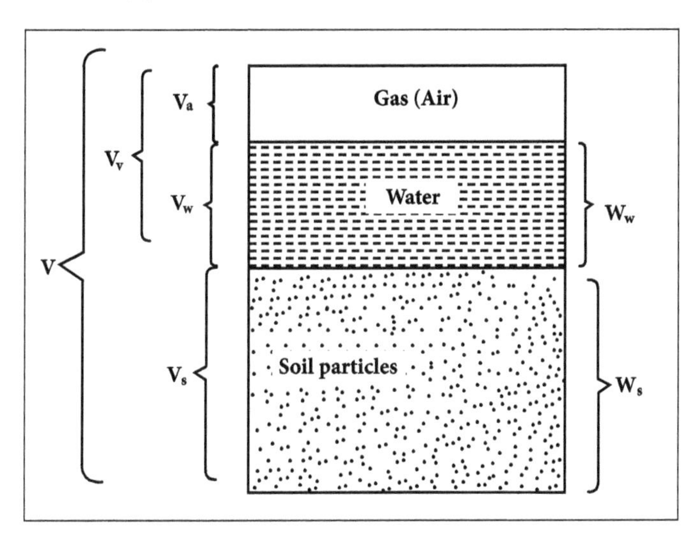

It should be noted that the value of porosity n is always less than 1.0. But, the value of void ratio e may be less, equal to or greater than 1.0. Further, if the weight of water in a wet soil sample is W_w and the dry weight of the sample is W_s, then the dry weight moisture fraction, W is expressed as,

$$W = \frac{W_w}{W_s} \qquad ...(2)$$

The bulk density γ_b of a soil mass is the total weight of the soil per unit bulk volume, i.e.,

$$\gamma_b = \frac{W_T}{V}$$

In which,

$$W_T = W_s + W_w$$

The specific weight (or the unit weight) of the solid particles is the ratio of dry weight of the soil particles W_s to the volume of the soil particles $V_{s,}$ i.e. W_s/V_s. Thus,

$$G_b\gamma_w = \frac{W_s}{V} \quad \text{i.e,} \quad V = \frac{W_s}{G_a\gamma_\omega}$$

And,

$$G_s\gamma_w = \frac{W_s}{V_s} \quad \text{i.e,} \quad V_s = \frac{W_s}{G_a\gamma_\omega}$$

Therefore,

$$\frac{V_s}{V} = \frac{G_b}{G_s} \qquad \ldots(3)$$

Here, γ_ω is the unit weight of water and G_b and G_s are, respectively, the bulk specific gravity of soil and the relative density of soil grains. Further,

$$1 - n = 1 - \frac{V_\upsilon}{V} = \frac{V - V_\upsilon}{V} = \frac{V_s}{V} = \frac{G_b}{G_s}$$

$$G_b = G_s(1-n) \qquad \ldots(4)$$

Also,

$$\omega = \frac{V_\omega}{V} = \frac{W_\omega/\gamma_\omega}{W_s/(G_b\gamma_\omega)} = G_b\frac{W_\omega}{W_s}$$

$$\omega = G_b W \qquad \ldots(5)$$

And,

$$\omega = G_s(1-n)W$$

Considering a soil of root-zone depth d and surface area A (i.e. bulk volume = Ad),

$$W_s = V_s G_s \gamma_\omega = Ad(1-n)G_s\gamma_\omega$$

Therefore, the dry weight moisture fraction, $W = \frac{W_\omega}{W_s}$

$$= \frac{V_\omega\gamma_\omega}{Ad(1-n)G_s\gamma_\omega}$$

Therefore, the volume of water in the root-zone soil,

$$V_\omega = W\ Ad(1-n)G_s \qquad \ldots(6)$$

This volume of water can also be expressed in terms of depth of water which would be obtained when this volume of water is spread over the soil surface area A.

Therefore, depth of water is, $d_\omega = \frac{V_\omega}{A}$

$$d_\omega = G_s(1-n)W_d \qquad ...(7)$$

or,

$$d_\omega = \omega\, d \qquad ...(8)$$

1.6 Canal Irrigation

The canals used for irrigation purpose are known as irrigation canals. Canals may be defined as artificial channels constructed on the ground to carry the water from one place to the other. The canals may be classified as alluvial or non-alluvial canal based on the nature of source of supply.

It may be termed as inundation or a permanent canal depending on how the water is fed from one system to another.

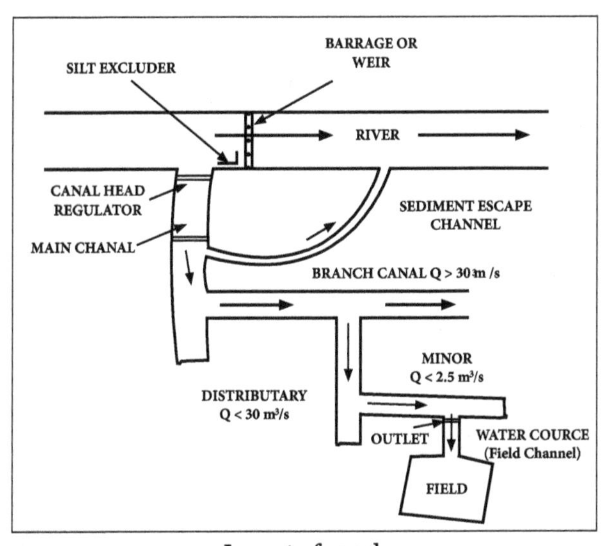

Layout of canal.

An irrigation canal system consists of canals of different sizes and capacities.

Based on different sizes and capacities the canals are also classified as:

- Main canal.
- Branch canal.

- Major distributary.
- Minor distributary.
- Watercourse.

1. Main Canal

Main Canal takes off directly from the upstream side of weir head works or dam. Usually no direct cultivation is proposed.

Main canal.

2. Branch Canal

All offtakes from main canal with head discharge of 14-15 cumecs and above are termed as branch canals. Acts as feeder channel for major distributaries.

Branch canal.

3. Major Distributary

All offtakes from main canal or branch canal with head discharge from 0.25 to 15 cumecs are called as the major distributaries.

4. Minor Distributary

All offtakes taking off from a major distributary getting discharge less than 0.25 cumec are termed as minor distributaries.

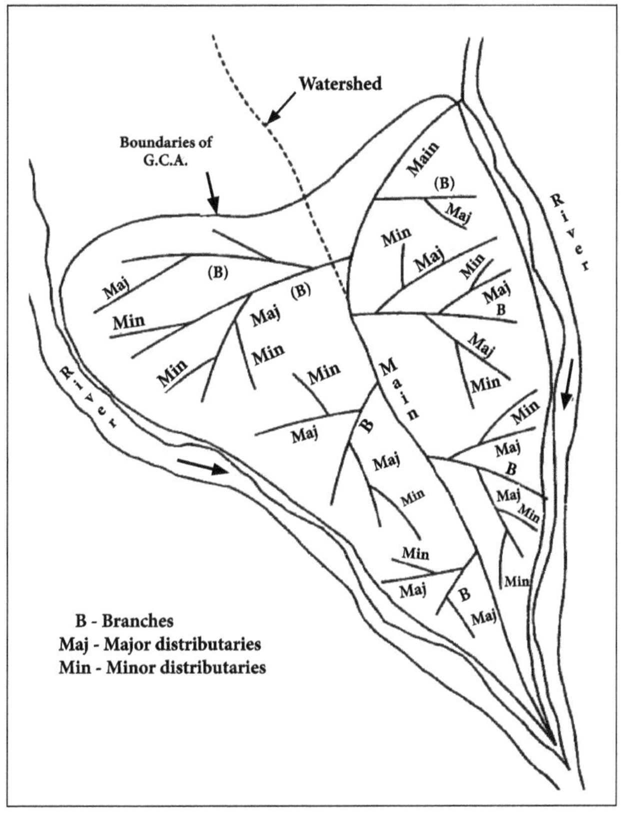

Distributary

5. Watercourse

Small channels which carry water from outlets of a major or minor distributary or a branch canal to the fields to be irrigated.

Classification of Canal Based on Supply

1. Alluvial Canal

A canal flowing through alluvium soil is called an alluvial canal. A canal flowing through those sediments transports some of this material along the flowing water. These canals take supplies from rivers which always carry sediments rolling on the bed or held in suspension, which is passed on to the off taking canals. If the velocity in a canal is very high, the suspension particles are not deposited, but when velocity is very low, the sediment held in suspension will get deposited.

2. Non Alluvial Canal

Non alluvial canals are canals that have been lined with some suitable material to provide a rigid bed banks so as to avoid the conflicts with the alluvial sides of a canal.

3. Inundations Canal

Those canals which depend for their supply on the periodical rise in canals are not always of the desired level. The water level of the river from which they are taken off. The supplies of these canals are not always of the assumed level. These canals are filled with water in Rainy season or in monsoon.

4. Permanent Canal

A canal is said to be a permanent canal when its source of supply is sufficiently well assured so as to warrant the construction of a regular grade channel supplied for the regulation and distribution. These canals are provided with a permanent masonry head works, distribution and regulator works. They are also constructed with engineering skills. The lining of the irrigation canal is also protective work because it helps in minimizing the chances of water- logging.

5. Feeder or Link Canal

Link canals supply water from a reservoir to another place wherefrom a given irrigation canal system is fed. These canals are used for diverting surplus water from one source to another.

6. Productive Canals

The canals, which indicate at the time of design and planning, that the total income will exceed the annual maintenance charges, are called productive canals.

Classification of Canals based on Alignment

1. Watershed Canal or Ridge Canal

The dividing ridge line between the catchment areas of two streams is called the watershed or ridge canal. Thus between two major streams, there is the main watershed, which divides the drainage area of the two streams.

The canal which is aligned along any natural watershed is called a watershed canal or a ridge canal. Aligning a canal on the ridge ensures gravity irrigation on both sides of the canal.

Since the drainage flows away from the ridge, drainage cannot cross a canal aligned on the ridge. Thus, canal aligned on the watershed saves cost of construction of cross-drainage works.

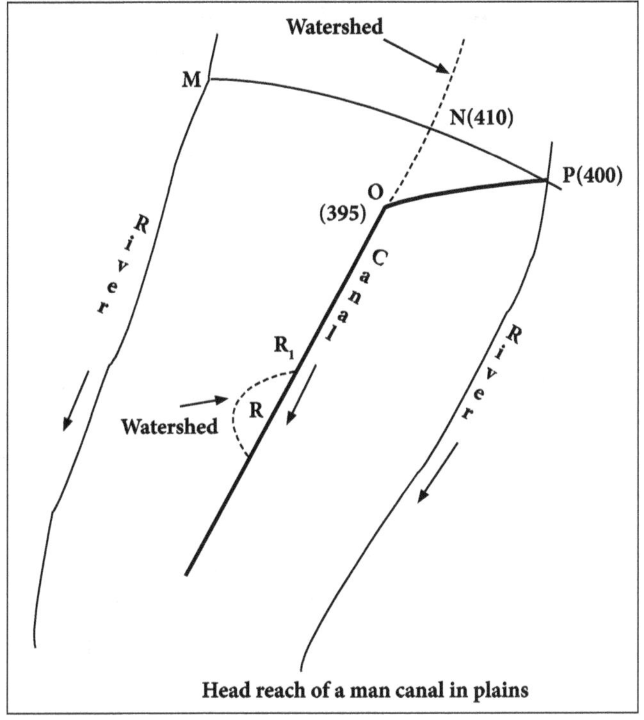

Ridge canal.

2. Contour Canal

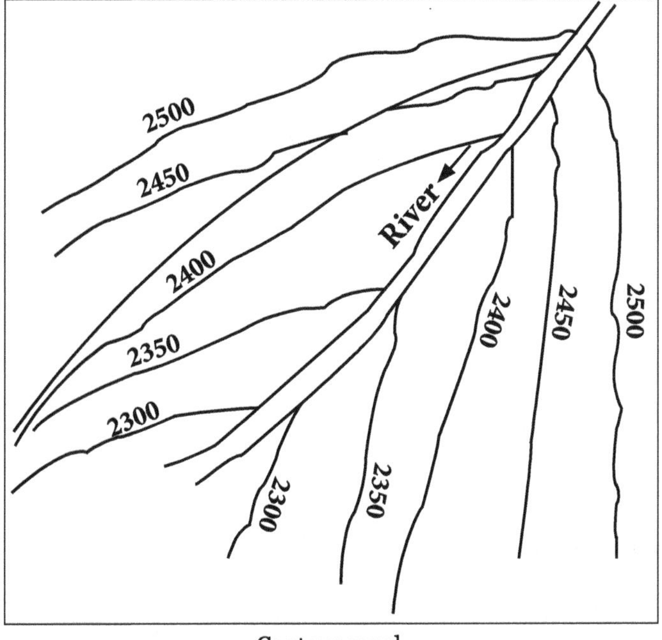

Contour canal.

Watershed canal along the ridge line are not found economical in hill areas. In hills, the river flows in the valley well below the watershed.

In fact, the ridge line may be hundreds of meters above the river, it therefore becomes virtually impossible to take the canal on top of such a higher ridge line. In such conditions, contour canals are usually constructed.

A contour canal irrigates only one side because the area on the other side is higher.

3. Side Slope Canal

A side slope canal is one which is aligned at right angles to the contours, i.e. along the side slopes.

Since such a canal runs parallel to the natural drainage flow, it usually does not intercept drainage channels, thus avoiding the construction of cross-drainage structures. It is a canal which is aligned roughly at right angle to contours of the country but not on watershed or valley.

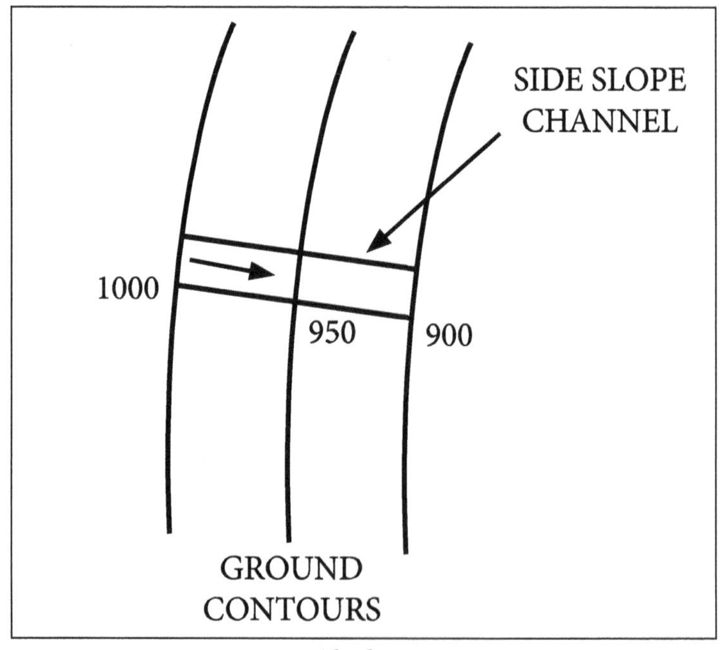

Side slope.

Classification of Canals (Based on Lining)

1. Unlined Canal

- Bed and banks made up of natural soil.
- Water velocities higher than 0.7 m/s are not tolerable.
- High seepage and conveyance water losses.
- Profuse growth of aquatic weeds reduces the flow.

Unlined canal.

2. Lined Canal

- Lining of impervious material on its bed and banks to prevent the seepage of water.
- Different types of lining used e.g. concrete, brick or burnt clay tile, boulder, etc.

Lined canal.

Canal Losses

Loss Due to Evaporation

As canal water is exposed to the atmosphere at the surface, loss due to evaporation is obvious. It is of course true that in most of the cases evaporation loss is not significant. It may range from 0.25 to 1% of the total canal discharge.

The rate of loss of water in the process of evaporation depends mainly on the following factors:

- Temperature of the region.
- Prevailing wind velocity of the region.

- Humidity.
- Area of water surface exposed to the atmosphere.

Generally it is considered that the rate of evaporation loss depends mainly on temperature. It is not hundred per cent correct. The rate of loss also equally depends on the velocity of wind which carries vapour from the water surface to the atmosphere.

Loss due to evaporation is more for shallow water depths. Many times it is observed that due to above mentioned factors the rate of loss due to evaporation does not differ much for day and night periods.

Thus it can be inferred that the loss due to evaporation is directly dependent on the climatic conditions of the region and hence cannot be checked. It also depends directly on the exposed area of the water surface and inversely on the depth of water in the channel.

Loss Due to Seepage

The water lost in seepage may find its way finally into the river valley on enters an aquifer where it can be utilized again. But many times the seepage water is not recoverable.

The loss due to seepage is the one which is most significant as far as irrigation water loss from a canal is concerned.

The seepage loss depends mainly on the following factors:

- Underground water table conditions.
- Porosity of the soil.
- Physical properties of the canal water for example its temperature and quantity of suspended load carried by the water (turbidity of water).
- Condition of the canal system.

Alignment of Canals

The total cost of a canal project depends upon the alignment. The alignment is the feasible path or route from a source location to the desired destination. A canal has to be aligned in such a way that it covers the entire area proposed to be irrigated with the shortest possible length and at the same time, its cost including the cost of cross-drainage works is a minimum.

A shorter length of canal ensures less loss of head due to friction and smaller loss of discharge due to seepage and evaporation, so that additional areas can be brought under cultivation. Canal alignment may be contour canal, side slope canal or ridge canal as per terrain of command area.

A contour canal irrigates only one side of the canal and it crosses a number of valleys, thus, it involves different types of cross-drainage works such as aqueducts, under tunnels, super passages, etc. A side slope canal is aligned at right angles to the contours of a country. A watershed or ridge canal irrigates the areas on both sides.

Cross-drainage works are completely eliminated in watershed and side slope canals. The main canal is generally carried on a contour alignment, until either it commands the full area to be irrigated or it attains the top of a watershed to become a watershed canal thereafter. Branch canals and distributaries take off from a canal from or near the points where the canal crosses the watershed.

The alignment of a canal is decided after a careful consideration of the economy. Several alignments between the source and the destination may be possible. An alignment mainly depends on the topography. Out of many alignments, few may not be feasible to construct due to construction-related problems. Canals are aligned as far as possible in partial cutting and partial filling.

Deep cutting or high embankments are generally avoided by suitable detouring after comparing the overall costs of the alternative alignments. Land cost varies with land use pattern, resettlement and rehabilitation cost, environmental cost and alignment of the canal, the cost of canal falls/drops/cross-drainage works varies with the type and size of structure. The maximization of economy is achievable by minimization of the total cost of canal route alignment considering all possible cost factors.

1.7 Design of Stable Channels using Kennedy's and Lacey's Theory

Kennedy's Silt Theory

Robert Kennedy was an engineer (1895) in Punjab Irrigation Department. He made extensive study in 22 irrigation canals and their distributaries in upper Bari Doab area of then Punjab, India. From the data he collected from those old stable canals, presented the following non-silting and non-scouring velocity V_c initially in fps unit as,

$$V_c = 0.84\psi^{0.64}$$

Converting this original equation is MKS unit system,

$$V_c = 0.547159\psi^{0.64}$$

i.e.,

$$V_c = 0.55\psi^{0.64}$$

Where,

V - The non-silting and non-scouring velocity in m/sec.

y - The depth in meter.

Kennedy claimed that sediment size plays an important role and so he defined a critical velocity ratio $(CVR) = m_r \frac{V}{V_c}$. Where V is the actual velocity.

$$V = m_r V_c$$

$$V = m_r \times 0.55\psi^{0.64}$$

He found $m_r > 1$ (i.e., 1.1 to 1.2) for sediment coarser than upper Bari Doab and $m_r < 1$ (0.8 to 0.9) for sediment finer than his canals. Actual velocity V is given by Ganguillet and Kuiters equation, i.e.,

$$V = \left[\frac{\left(23 + \frac{0.00155}{S_b} + \frac{1}{n_r} \right)}{1 + \left(23 + \frac{0.00155}{S_b} \right) \frac{n_r}{\sqrt{R}}} \right] \sqrt{Rs_b} \quad ...(1)$$

Where n_r is called Kutter's n_r.

Kennedy claimed that due to generation of eddies in the canal bed and its rise to the surface, flowing water gets the silt supporting power. These eddies are generated due to friction of flowing water with channel surface. Vertical component of such eddies try to cause the sediment to move up while weight of sediment acts downward and thus, the sediment is kept in suspension and silting is avoided.

Steps Required to Design Canal by Kennedy's Theory

The known values are discharge Q, bed slope s_b (decided by topography), CVR m_r and Kutter's n_r. Trial and error method has to be adopted to determine the design values of y and B. Side slope 1 V is assumed (if not given). i.e. z varies from 0.5 to 2 for hard to loose soil. It is based on angle of repose of earth.

Steps:

- Assume a trial value of depth y.
- Find the velocity by the equation $V = 0.55 m_r \psi^{0.64}$.
- Find flow area $A = \frac{Q}{V}$.

- Find B from $A-(B+z\psi)\psi$ where A is known from step (3), z is assumed or given. ψ is assumed in step (1).
- Find $P=B+2y\sqrt{1+z,^2}$.
- Find $R=\frac{A}{P}$.
- Now find velocity V by Kona's equation, i.e., Equation (1).
- If the velocity obtained in step (2) and in step (7) are not almost equal. Assume second trial values of depth y and repeat steps (1) to (8).
- Repeat this process, until velocity in step (2) is equal to velocity in step (7). The value of y at which the two velocities are almost same, is the required depth of flow.
- Find B when y is known.
- Assume a reasonable free board (FB). Usually 0.6 when Q < 10 m^3/sec and FB > 0.75 when Q > 10 m^3/sec.

This trial and error method is very tedious and assumption of first trial values ay is very difficult. The design of canal by Kennedy's theory can be designed from Garret's diagram which provide a graphical solution of Kennedy's equation and Cianguillet–Kutter's equation with interpolation.

Drawbacks of Kennedy's Theory:

- In his equation $V_c = Ky_n$ the values of K vary with sediment size. K = 0.55 and n=0.64 are only applicable to Bari Doab canal sediment, hence, these values cannot safely be applied to canal of different soil and sediment sizes.
- Kennedy did not give any equation or method of measurement of m, (CVR).
- Kennedy equation shows that velocity is only a function of depth y. Shape of channel, width, roughness, slope are not at all considered in assuming the velocity.
- Assumption of first assumed value of depth y to initiate trial and error method is very difficult. Some approximate value based on discharge, slope, etc. should have been given to save the computational time.
- His regime velocity. i.e. non-silting and non-scouring velocity did not consider the sediment load as variable.
- Gunguillet–Kutter's equation to check the mean velocity of flow with sediment has limitation as the equation is for non-erodible canal.
- He did not give any equation of slope.
- Complex silt carrying phenomenon cannot be represented by only one factor like in (CVR).

- y/B ratio which is dependent on Q was not noticed by Kennedy.
- It is entirely empirical and also requires rigorous trial and error which is tedious and time consuming.

Lacey's Theory: Concept, Equations and Limitations

Concept

Taking lead from Kennedy theory Mr. Gerald Lacey undertook detailed study to evolve more scientific method of designing the irrigation channels on alluvial soils. He presented revised version of his study in 1939 which is popularly known as Lacey's theory.

In this theory, Lacey explained in detail concept of regime conditions and rugosity coefficient.

It may be seen that for a channel to achieve regime condition following three conditions have to be fulfilled:

- Channel should flow uniformly in "incoherent unlimited alluvium" of same character as that transported by the water.
- Silt grade and silt charge should be constant.
- Discharge should be constant.

These conditions are very rarely achieved and are very difficult to maintain in practice. According to Lacey's concept, regime conditions may be subdivided as initial and final.

In rivers, achievement of initial or final regime is practically impossible. Only in bank full stage or high floods the river can be considered to achieve temporary or quasi-regime. Recognition of this fact can be utilized to deal with the issues concerning the scour and floods.

Lacey also state that the silt is kept in suspension solely by force of eddies. Lacey adds that eddies are not generated on the bed only but at all points on the wet perimeter. The force of eddies may be taken normal to the sides.

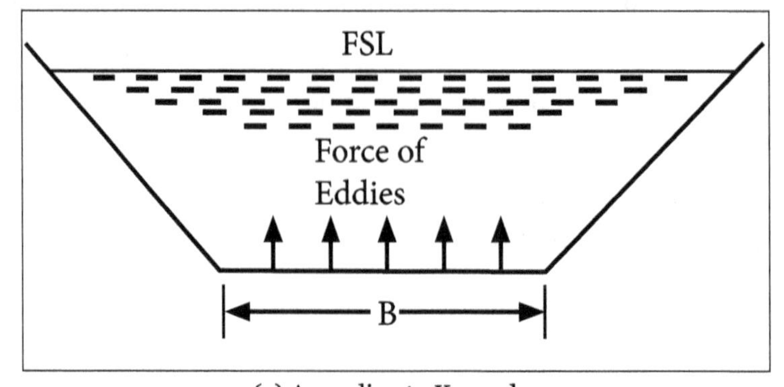

(a) According to Kennedy.

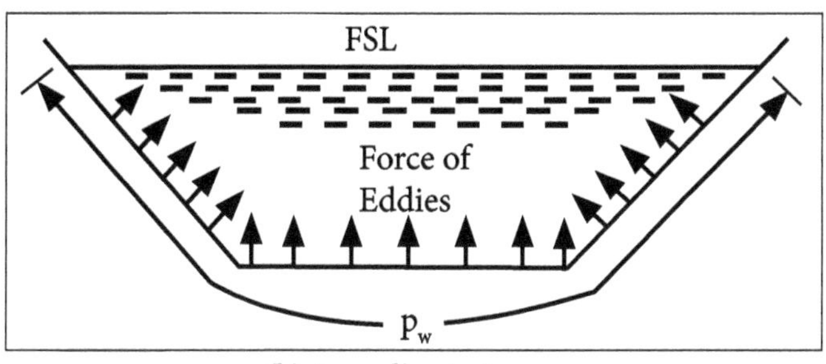

(b) According to Lacey.

The vertical components of forces due to eddies are responsible for keeping the silt in suspension. Unlike Kennedy, the Lacey takes hydraulic mean radius (R) as a variable rather than depth (D).

As wide channels are concerned there is hardly any difference between R and D. When the channel section is semi-circular there is no base width and sides actually.

Consequently amount of silt transported is not dependent on the base width of a channel only. On the basis of arguments Lacey plotted a graph between regime velocity (V) and hydraulic mean radius (R) and gave the relationship,

$$V = K.R_1 \quad ...(1)$$

Where K is a constant.

Power of R is a fixed number and it does not need alteration to suit different conditions. Lacey recognized the importance of silt grade in the problem and introduced a concept of function 'f' known as silt factor. He adjusted the values such that it also came under square root sign. It gave scalar conception.

Equation (1) is thus modified as,

$$V = K.\sqrt{f}.R \quad ...(2)$$

Kennedy's general equation is,

$$V = c.m.Dn \quad ...(3)$$

Comparing equations (2) and (3),

$$f = m^2$$

From equation (2) it can be seen that in regime channels if the mean velocity is same then hydraulic mean radius varies inversely with the silt factor.

Lacey takes silt as a standard silt when silt factor is unity for that silt. He further states that standard silt is sandy silt within a regime channel with hydraulic mean radius equal to one metre.

Lacey's Regime Equations

After study and plotting of large data to justify this theory Lacey gave three fundamental equations from which other equations have been derived for design of irrigation channels.

The three fundamental equations are,

$$V = 0.639\sqrt{fR} \qquad ...(1)$$

Where V is regime velocity in m/sec,

$$Af^2 = 141.24\ V^5 \qquad ...(2)$$

$$V = 10.8\ R^{2/3}\ S^{1/3} \qquad ...(3)$$

Where S is the slope of water surface.

Equation (3) is termed as the regime flow equation and is of great practical importance.

It may be seen that the equation does not contain the term of rugosity coefficient. It is necessary to know the value of coefficient of rugosity, the selection of which remains a matter of experience and many not be reliable especially in case of rivers in floods.

Considering that at high flood river flows in quasi-regime the regime flow equation (3) given above may be adopted though it may have some exceptions. Important equations given by Lacey in his theory are summarized below.

The first three equation are called fundamental equations on the basis of which other equations have been developed,

$$V = 0.639\ \sqrt{fR} \qquad ...(4)$$

$$Af^2 = 141.24\ V^6 \qquad ...(5)$$

$$V = 10.8 R^{2/3}\ S^{1/2} \qquad ...(6)$$

$$V = \frac{1}{N_o} R^{3/4} S^{1/2} \qquad ...(7)$$

$$N_a = 0.0225 f^{1/4} \qquad ...(8)$$

$$V = 0.4382\left(Q.F^2\right)^{1/6} \qquad ...(9)$$

$$R = 2.46\frac{V^2}{f} \quad ...(10)$$

$$P\omega = 4.825\sqrt{Q} \quad ...(11)$$

$$S = \frac{f^{5/3}}{3316Q^{1/6}} \quad ...(12)$$

$$A = 2.28\frac{Q^{5/6}}{f^{1/3}} \quad ...(13)$$

$$F = 1.76\sqrt{m_r} \quad ...(14)$$

$$R = 1.35\left(\frac{q^2}{f}\right)^{1/3} \quad ...(15)$$

Although all the above equations are given for regime conditions and are normally for canals in alluvium as the river attains quasi-regime conditions equations (6), (11) and (15) are very useful in calculating flood discharge, waterway needed during floods and scour depth during floods respectively.

When equation (15) is applied to rivers in flood, the value of R gives normal depth of scour. This formula is very useful in determining the levels of foundations, vertical cut-offs etc. This formula is popularly called Lacey's scour depth formula.

Limitations of Lacey's Theory

- The Lacey's work is based on the field observations and empirically derived equations and therefore it cannot be said to be theory in the strict sense.
- Regime equations in their derived form cannot be applied universally. This is because it hold good mostly for the regions whose data was taken for study.
- Like Kennedy's theory even though perfect definition of silt grade and silt charge is not given in most of the equations they are based on the silt factor 'f'.
- In practice regime condition stated by Lacey is rarely achieved and that too after a long period.
- The field observations show limited acceptance of the concept of semi-elliptical section of a regime channel.
- Complex phenomenon of sediment concentration and transport has not been scientifically considered.

Design of Irrigation Channels Making use of Lacey Theory

The full supply discharge for any canal is always fixed before starting a design. The value of 'f' for particular site can be calculated using equation (11) or if CVR is given then $f = m^2$.

Thus when Q and f are known design can be done in the following steps:

- Find out F using equation (6),

  $V = 0.4382\left(Q.f^2\right)^{1/6.}$

- Calculate value of R using equation (7),

  $R = 2.46v^2 / f$

- Calculate wetted perimeter P_w using Lacey's regime perimeter equation $P_w = 4.825Q^{1/2}$.

- Calculate cross sectional area A from equation Q = AV.

- Assuming side slopes, calculate the full supply depth from A, P_w and R.

- Calculate the longitudinal slope of the channel using equation (9).

Problem

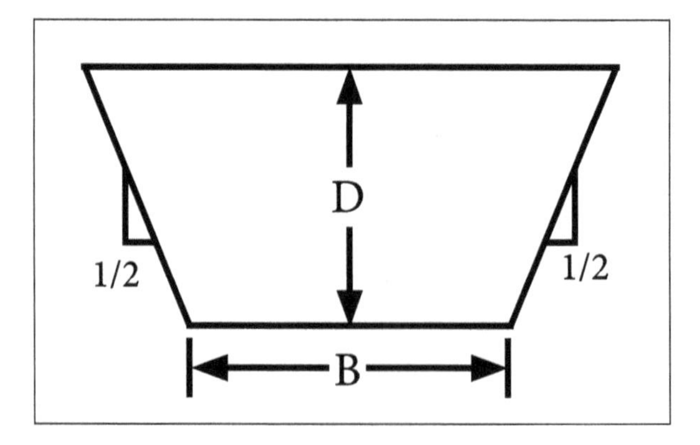

Let us design an irrigation canal by Lacey's theory for the following data,

$FSD = 14m^2 / sec$

$f = 1$

Side slopes = 1/2 : 1 (Horizontal : Vertical)

Coefficient of rugosity, N = 0.0225

Solution:

Given:

$$FSD = 14m^2 / sec$$

$$f = 1$$

Side slopes = 1/2 : 1 (Horizontal : Vertical)

Coefficient of rugosity, N = 0.0225

Mean velocity $v = 0.4382(14 \times 1^2)^{1/6} = 0.68m/sec$

$$R = 2.46\frac{(0.68)^2}{1} = 1.14m$$

Hydraulic mean radius.

Wetted perimeter $P_w = 4.25 \times (14)^{1/2} = 18.06m$

Cross sectional area, $A = R \times P_w = 1.14 \times 18.06 = 20.55\ m^2$

From figure,

$$A = BD + \frac{D^2}{2} \qquad ...(1)$$

$$P_w = B + \sqrt{5}D \qquad ...(2)$$

Putting values of A and P_ω in equations (1) and (2) respectively,

$$20.55 = BD + D^2/2$$

$$18.06 = B + \sqrt{5}D \qquad ...(3)$$

From equation (3),

$$B = 18.06 - \sqrt{5}\ D$$

Now putting value of B in equation (1),

$$20.55(18.06 - \sqrt{D})D + D^2 / 2$$

(or),

$$20.55 = 18.06D - 1.73D^2$$

or,

$$1.73D \ -18.06D + 20.55 = 0$$

It is a quadratic equation in D. Hence, solving for D.

$$\frac{18.06 \pm \sqrt{(18.06) \ -4 \times 1.73 \times 20.55}}{2 \times 1.73}$$

$$\frac{18.06 \pm 13.56}{3.46}$$

$$D \ 9.12 \text{ and } 1.13\,\text{m}$$

Taking,

$$D = 9.13\,\text{m}$$

B = 18.06 – 2.23 x 9.13 =- 2.14 which is absurd.

Hence, taking D = 1.3 m

$$B = 18.06 - 2.23 \times 1.3 = 15.15\,\text{m}$$

By using the values of B and D hydraulic mean radius can be calculated again to compare it with the values already calculated above using Lacey's equation,

$$R = 2.46 \frac{(0.68)^2}{1}$$

$$R = \frac{A}{P} = \frac{15.15 \times 1.3 + (1.3)^2 / 2}{15.15 + \sqrt{5} \times 1.3} = \frac{20.54}{18.05} = 1.14\,\text{m}$$

Now slope can be determined from formula,

$$S = \frac{f^{5/3}}{3316 \times Q^{1/6}} = \frac{1^{5/3}}{3316 \times (14)^{1/6}} = \frac{1}{3316 \times 1.553} = \frac{1}{5160}$$

Result:

FS depth 1.3 m, bed width 15.15 m and bed slope 1 in 5160.

1.7.1 Garret's Diagram, Cross Section of Irrigation Canals

Lot of mathematical calculations is required in designing irrigation channels by the use of Kennedy's method. To save mathematical calculations, graphical solution of Kennedy's and Kutter's equations, was evolved by Garret.

The original diagrams given by him were in F.P.S. system, but here they have been changed into M.K.S./S.I. system. The procedure adopted for design of irrigation channels using Garret's diagrams is explained below:

- The discharge, bed slope, rugosity coefficient, value of C.V.R. is given for the channel to be designed.
- Find out the point of intersection of the given slope line and discharge curve. At this point of intersection, draw a vertical line intersecting the various bed width curves.
- For different bed widths (B), the corresponding values of water depth (y) and critical velocity (V_0) can be read on the right hand ordinate. Each such pair of bed width (B) and depth (y) will satisfy Kutter's equation and is capable of carrying the required discharge at the given slope and rugosity coefficient. Choose one such pair and determine the actual velocity of flow.
- Determine the critical velocity ratio (V/V_0) taking V as calculated and VQ as read.
- (V_0) If the value of C.V.R. is not the same as given in question, repeat the procedure with other pairs of B and y.

Cross-Section of Irrigation Canal

When the full supply level of an irrigation canal is lower than the surrounding ground level (Figure(a)), the canal is said to be in cutting. If the canal bed is at or higher than the surrounding ground level the canal is in filling.

When the surrounding ground level is in between the full supply level and the bed level of the canal (Figure (c)), the canal is partly in cutting and partly in filling and such a canal, usually, results in balanced earthwork.

For the purpose of proper inspection and maintenance of irrigation canals, service roads are provided on one or both sides of the canal. These service roads can be part of the bank embankment as shown in Figure.

These service roads are about 6 m wide and may or may not be metaled. If these service roads are likely to meet the communication needs of the local people, they will be much wider and also metaled. Between the canal road and the canal, a 'dowla' of height 0.5 m and top width 0.5 m is also provided.

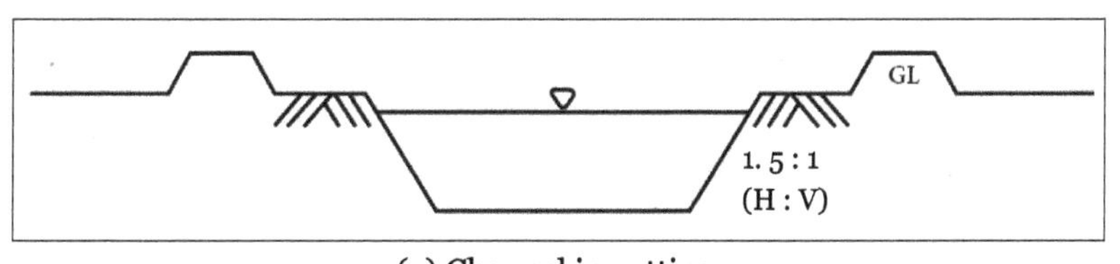

(a) Channel in cutting.

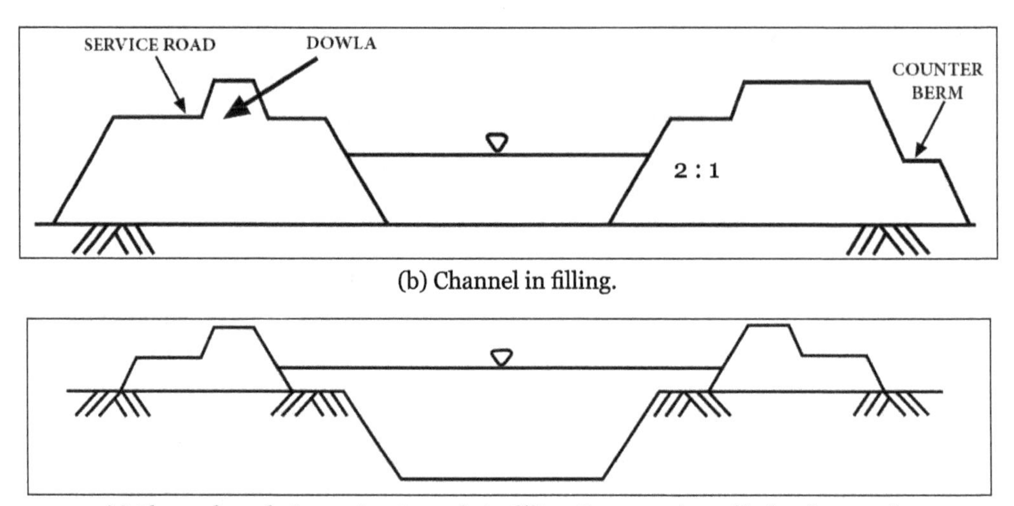

(b) Channel in filling.

(c) Channel partly in cutting & partly in filling Cross section of irrigation canals.

To prevent sloughing of inner surface and canal banks, a narrow horizontal strip is provided on the inner sloping surface of the bank. This strip is known as berm. For stability of banks, side slopes of an unlined channel should, obviously, be flatter than the angle of repose of the saturated bank soil. Generally, it varies from 1 : 1 to about 2(H) : 1(V).

Further, the bank dimensions should be such that there is a minimum cover of 0.5 m above the saturation line which, for small embankments, may be approximated by a straight line drawn at a slope ranging from 4H:1V (for relatively impermeable material) to 10H:1V (for porous sand and gravel) from the point where full supply level meets the bank. Cover over saturation line can be increased by providing a horizontal strip similar to berm on the outer slope of bank in which case this strips is known as counter berm (Figure(b)).

It also helps in collecting rain water and disposing of the same without letting rainwater make a continuous gully on the outer bank. The calculations for the design of an irrigation canal are carried out in a tabular form and the table of these computations is referred to as the 'schedule of area statistics and channel dimensions'. The computations start from the tail end and the design is usually carried out at every kilometer of the channel downstream of the head of the channel.

1.8 Lining of Irrigation Canals: Advantages, Economics and Various Types

Lining the Canals

It is always assumed that seepage losses would be reduced from the theoretical calculations as between 3.7-1.8 cumec per million square meter (cumec/mm^2) for an unlined canal in the sandy or clayey loams to 0.6 cumec/mm^2 for a lined canal.

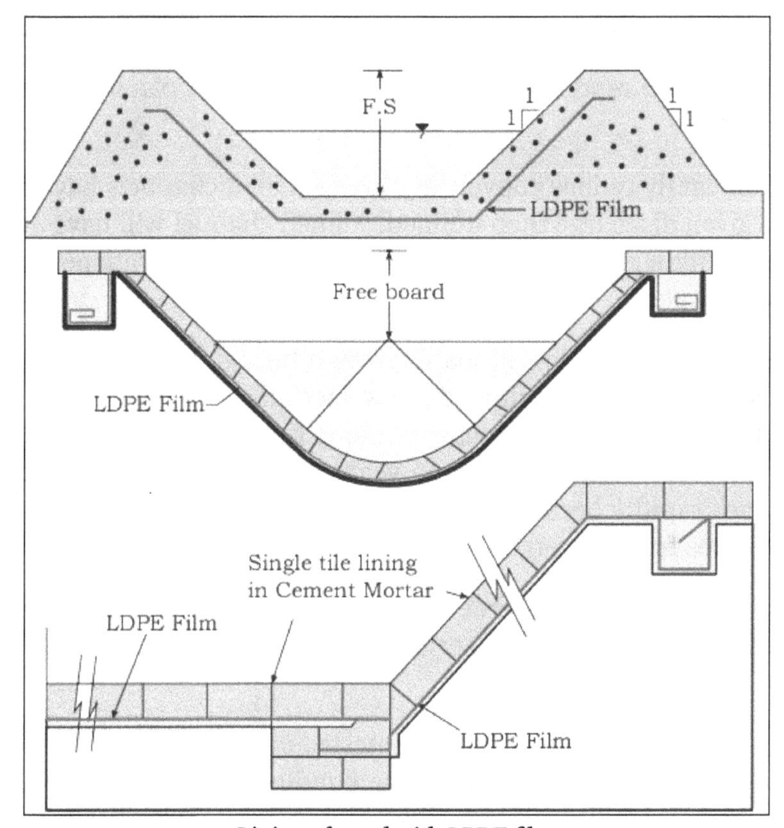

Lining of canal with LPDE film.

In 1988 a field study of the performance of lined distributaries were losing 3.5 cumec/mm^2 and water courses were losing 3.7 cumec/mm^2. Old earthen irrigation channels in permeable soils can lose a lot of water through seepage.

Large losses through the bed and sides of canal lead to low conveyance efficiency that is, (the ratio of water reaching farm turnouts to that released at the source of supply from a river or reservoir).

Earthen canals also get clogged up with weeds which reduce the water-carrying capacity. These two factors combined to be disadvantageous to the tail end farmers. Therefore Unlined canals are inefficient inadequate from the point of view of equitable performance.

In Punjab, the expected saving by brick lining is of the order of 20%. The brick linings have 25 years of life. Total losses from unlined watercourses are known to be more than those from the main system, but they don't get the same attention during a lining programme.

Lining programmes are divided into main system lining and watercourse lining. The main system canals (main, distributaries and minors) are large channels supplying several watercourses.

A typical value for the seepage rate in an unlined channel in clayey loam is 1.8 m^3/s.

Per mm^2 and through a rendered brick line water course or canal reduces to 0.1 m^3/s per mm^2.

Lining can significantly reduce conveyance losses. Lined channels have a smaller surface area for a given discharge than unlined. A lined channel will have 40% of the unlined surface area for a given discharge. Even at the same loss rate per unit area there will be a saving in water.

When estimating the reduction in losses from a lining programme, this should be based on the combination of a reduced cross-section and a reduced seepage rate per unit area. In the Indian Punjab, measurements on lined distributaries and watercourses between four months and seven years old showed that seepage rates from the distributaries rapidly became comparable to seepage rates from unlined canals while seepage rates from watercourses were highly variable.

Lining of the distributary canal seemed to have had a beneficial effect on the equity of supply between watercourses. However the effects of lining watercourses is still to be established.

A sample of 15 out of 130 watercourses were tested on the 30,000 mudki subsystem of the Sirhind Feeder in Ferozpur district using ponding tests and inflow-outflow methods. The mean and variability of seepage losses increased dramatically for lining more than four years old.

Some lining older than four years performed as well as new linings, with losses as low as 0.4 m^3/s per mm^2 but others has losses of up to 11.5 m^3/s per mm^2. Conveyance losses were significantly greater than seepage losses alone. The variability of conveyance losses was observed to be related to the condition of the channels.

Losses from raised watercourses with cracked or broken linings appeared as surface leakage causing water logging of adjacent fields and localized crop damage. This is due to poor quality control during construction particularly earth compaction behind sidewalls and a lack of subsequent maintenance.

The design life of concrete and brick lined channels is generally assumed to be 25-50 years. Major repairs of lined channels are sometimes required within a few years of construction.

At the Kraseio Scheme in Thailand completed in 1981, long lengths of the concrete lined main canals have been replaced each year. In one 26-year lining test, a complete repair of the drained test channel was needed every 22 months. There is no watertight case for or against lining.

If lining goes ahead, a high standard of construction is essential especially of water courses which must withstand a great deal of wear and tear. Without adequate supervision poor construction of channels will lead to reduced life and higher maintenance costs.

Earthen watercourse in the Bikaner area of Rajasthan discharge an average 28.3 l/s but begin to seep and leak badly after little more than a year. These cracks and the slow movement of shallow water favors development of thick aquatic weeds that encourages the drying and the cracking process and structurally weaken the banks. This adds significantly to the cost of maintenance.

The cracks opened in dry periods do not close fully when saturated by water flows and losses can be up to 25% of the water diverted into the system. The cycle of swelling, heaving, shrinkage and settlement leads to progressive bank deterioration. Shear strength of clay depends on cohesion between particles.

In a newly-formed compacted clay masses, the inter particle cohesion is high. On first drying, the cracks appear and close up again on wetting but do not regain their original inter particle cohesion.

A reduction in shear strength occurs after a few drying and wetting cycles. Reinforced concrete lining would reduce the seepage loss drastically and has lower operation and maintenance costs.

Advantages of Canal Lining

- It reduces the loss of water due to seepage and hence the duty is enhanced.
- It controls the water logging and hence the bad effects of water-logging are eliminated.
- It provides smooth surface and hence the velocity of flow can be increased.
- Due to the increased velocity the discharge capacity of a canal is also increased.
- Due to the increased velocity, the evaporation loss also be reduced.
- It eliminates the effect of scouring in the canal bed.
- The increased velocity eliminates the possibility of silting in the canal bed.
- It controls the growth of weeds along the canal sides and bed.
- It provides the stable section of the canal.
- It reduces the requirement of land width for the canal, because smaller section of the canal can produce greater discharge.
- It prevents the sub-soil salt to come in contact with the canal water.
- It reduces the maintenance cost for the canals.

1.8.1 Economics of Canal Lining, Various Types of Lining

The economic viability of lining of a canal is decided on the basis of the ratio of additional benefits derived from the lining to additional cost incurred on account of lining. The ratio is worked out as follows:

Let C = cost of lining in Rs/sq. meter including the additional cost of dressing the banks for lining and accounting for the saving, if any, resulting from the smaller cross-sections and hence, smaller area of land, quantity of earth work and structures required for the lined sections.

This saving will be available on new canals excavated to have lined cross-section right from the beginning, but not on lining of the existing unlined canals.

s and S = Seepage losses in unlined and lined canals, respectively, in cubic meter per square meter of wetted surface per day of 24 hrs.

p and P = Wetted perimeter in meters of unlined and lined sections.

T = Total perimeter of lining in meters.

d = Number of running days of canal per year.

W = Value of water saved in rupees per cubic meter.

L = Length of the canal in meter.

y = Life of the canal in years.

M = Annual saving in rupees in operation and maintenance due to lining, taking into account the maintenance expenses on lining itself.

B = Annual estimated value in rupees of other benefits for the length of canal under consideration. This will include prevention of waterlogging, reduced cost of drainage for adjoining lands, reduced risk of breach and so on.

The annual value of water lost by seepage from the unlined section = pL sdW rupees.

The annual saving in value of water otherwise lost by seepage,

$$= (pLsdW - PL\,SdW) \text{ rupees}$$

$$= \{LdW\,(ps - PS)\} \text{ rupees}$$

Total annual benefits resulting from the lining of canal,

$$Bt = \{LdW\,(ps - PS) + B + M\} \text{ rupees}$$

Additional capital expenditure on construction of lined canal

$$= TLC \text{ rupees}$$

If the prevalent rate of interest is x percent per year, the annual installment a (rupees) required to be deposited each year (at its beginning) for a number of y years to amount to TLC plus its interest at the end of y years is determined by the following equation,

$$TLC\left(1+\frac{x}{100}\right)^{y}=a\left\{\left(1+\frac{x}{100}\right)^{y}=a\left(1+\frac{x}{100}\right)^{y-1}+\left(1+\frac{x}{100}\right)^{y-2}+\ldots+\left(1+\frac{x}{100}\right)^{1}\right\}$$

$$=\frac{a\left(1+\frac{x}{100}\right)\left\{\left(1+\frac{x}{100}\right)^{y}-1\right\}}{\left\{\left(1+\frac{x}{100}\right)-1\right\}}$$

$$a=\frac{(TCL)\left(\frac{x}{100}\right)\left(\frac{100+x}{100}\right)^{y-1}}{\left(\frac{100}{100}+x\right)^{y}-1}$$

For lining to be economically feasible, the value of a should be less than the annual benefit B_t i.e., the ratio B_t/a should be greater than unity. If b/y, z are specified the equation can be solved explicitly for y and b. The cost of materials used in lining a channel can be specified in terms of the value of material used. This may be expressed as,

$$AR^{2/3}=\frac{\left(by+my^{2}\right)^{5/3}}{\left[b+2y\sqrt{1+m^{2}}\right]^{2/3}}-\frac{Q_n}{\sqrt{S_o}}$$

Solve the above equation for y,

$$y=\frac{\left[b/y+2\sqrt{1+m^{2}}\right]^{1/4}}{(b/y+m)^{5/8}}\left(Q_n/\sqrt{S_o}\right)^{3/8}$$

Cost of bed material $C_b = \mu_B t_b(b + 2b') = Bb + k$ Per unit length

Cost of side material $C_s=\mu_s t_s(2E+E')=2\Gamma\left[(y+F_B)\sqrt{1+m^{2}}\right]$

Therefore,

$$C=C_b+C_s=bB+k+2\Gamma\left[(y+F_B)\sqrt{1+m^{2}}\right]$$

Where,

C = Total material cost per unit length.

C_b = Material cost per channel base per unit length.

C_s = Material cost of sides per unit channel length.

b' = Bottom corner width.

t_b = Thickness of the base material.

t = Channel side lining thickness.

μB = Cost of base lining material per unit volume.

μ_s = Cost of side lining material per unit volume.

B = Base material cost for specified thickness per unit area.

r = Cost of side lining material for specified thickness per unit area.

F_B = Vertical free board requirement.

E = Wetted length of the side.

E' = Side length of the free board.

Various Types of Lining

The following are the different types of linings which are generally recommended according to the various site conditions:

- Cement concrete lining.
- Pre-cast concrete lining.
- Cement mortar lining.
- Lime concrete lining.
- Brick lining.
- Boulder lining.
- Shot crete lining.
- Asphalt lining.
- Bentonite and clay lining.
- Soil-cement lining.

1. Cement Concrete Lining

This lining is recommended for the canal in full banking. The cement concrete lining (cast in-situ) is widely accepted as the best impervious lining. It can resist the effect of scouring and erosion very efficiently. The velocity of flow may be kept above 2.5 m/sec.

It can eliminate completely growth of weeds. The lining is done by the following steps,

- Preparation of sub-grade: The sub-grade is prepared by ramming the surface properly with a layer of sand (about 15 cm). Then, a slurry of cement and sand (1:3) is spread uniformly over the prepared bed.
- Laying of concrete: The cement concrete of grade M15 is spread uniformly according to the desired thickness (generally, the thickness varies from 100 mm to 150 mm). After laying, the concrete is tapped gently until the slurry comes on the top. The curing is done for two weeks.

As the concrete is liable to get damaged by the change of temperature, the expansion joints are provided at appropriate places. Normally no reinforcement is required for this cement concrete. But in special cases, a network of 6 mm diameter rods may be provided with spacing 10 cm centre to centre.

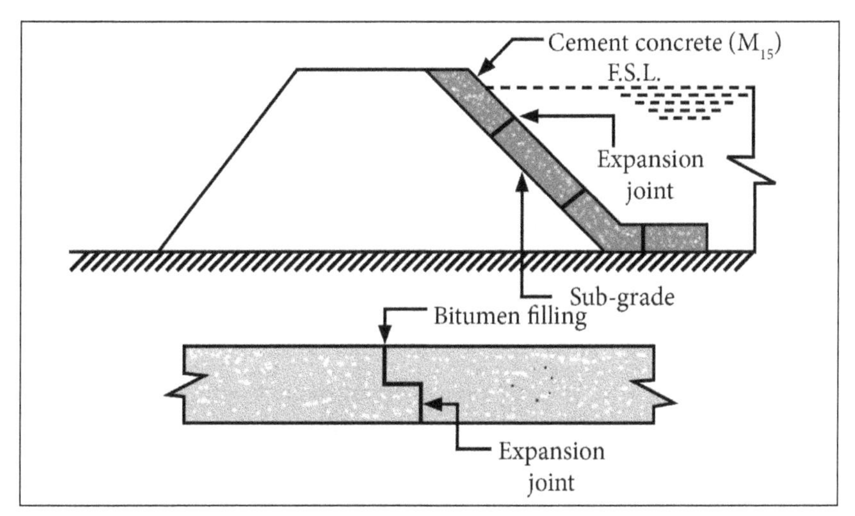

Cement concrete lining.

2. Pre-Cast Concrete Lining

This lining is recommended for the canal in full banking. It consists of pre-cast concrete slabs of size 60 cm x 60 cm x 5 cm which are set along the canal bank and bed with cement mortar (1:6). A network of 6 mm diameter rod is provided in the slab with spacing 10 cm centre of centre. The proportion of the concrete is recommended as 1:2:4.

Rebates are provided on all the four sides of the slab so that proper joints may be obtained when they are placed side by side. The joints are finished with cement mortar (1:3). Expansion joints are provided at a suitable interval. The slabs are set in the following sequence,

- The sub-grade is prepared by properly ramming the soil with a layer of sand. The bed is leveled so that the slabs can be placed easily.

- The slabs are stacked as per estimate along the course of the canal. The slabs are placed with cement mortar (1:6) by setting the rebates properly. The joints are finished with cement mortar (1:3).
- The curing is done for a week.

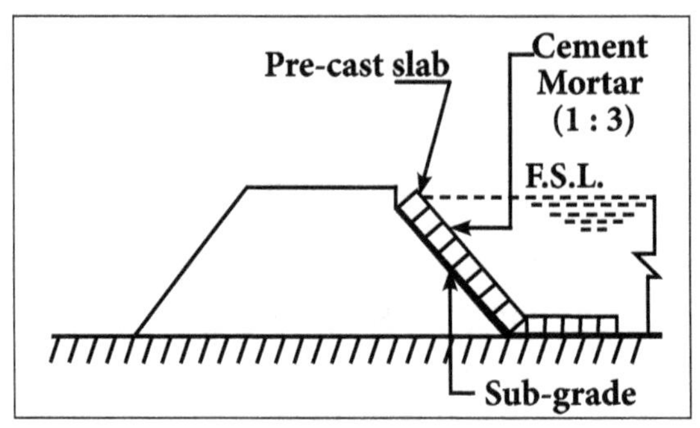

Pre-cast concrete lining.

3. Cement Mortar Lining

This type of lining is recommended for the canal fully in cutting where hard soil or clayey soil is available. The thickness of the cement mortar (1:4) is generally 2.5 cm. The sub-grade is prepared by ramming the soil after cutting. Then, over the compacted sub-grade, the cement mortar is laid uniformly and the surface is finished with neat cement polish. This lining is impervious, but is not durable. The curing should be done properly.

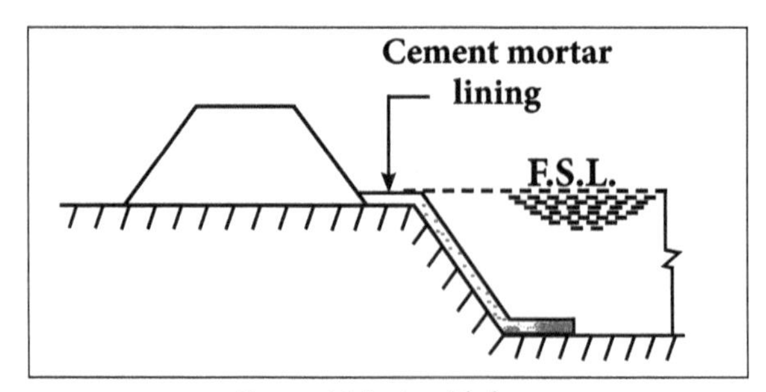

Cement Mortar Lining.

4. Lime Concrete Lining:

When hydraulic lime, surki and brick ballast are available in plenty along the course of the canal or in the vicinity of the irrigation project, then the lining of the canal may be made by the lime concrete of proportion 1:1:6. The procedure of laying this concrete is same as that of the cement concrete lining. Here, the thickness of concrete varies from 150 mm to 225 mm and the curing should be done for longer period. This lining is less durable than the cement concrete lining. However, it is recommended because of the availability of the materials and also because of the economics.

5. Brick Lining

This lining is prepared by the double layer brick flat soling laid with cement mortar (1:6) over the compacted subgrade. The first class bricks should be recommended for the work. The surface of the lining is finished with cement plaster (1:3). The curing should be done perfectly. This lining is always preferred for the following reasons:

- This lining is economical.
- Work can be done very quickly.
- Expansion joints are not required.
- Repair works can be done easily.
- Bricks can be manufactured from the excavated earth near the site.

However this lining has certain disadvantages:

- It is not completely impervious.
- It has low resistance against erosion.
- It is not so much durable.

6. Boulder Lining

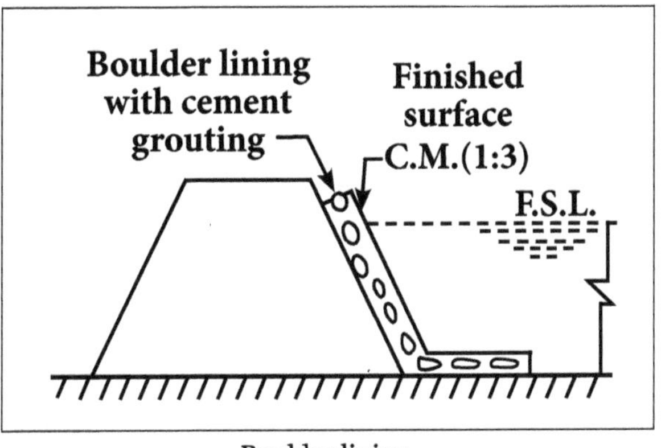

Boulder lining.

In hilly areas where the boulders are available in plenty, this type of lining is generally recommended. The boulders are laid in single or double layer maintaining the slope of the banks and the bed level of the canal. The joints of the boulders are grouted with cement mortar (1:6). The surface is finished with cement mortar (1:3).

Curing is necessary in this lining too. This lining is very durable and impervious. But the transporting cost of the material is very high. So, it cannot be recommended for all cases.

7. Shot Crete Lining

In this system, the cement mortar (1:4) is directly applied on the sub-grade by an equipment known as cement gun. The mortar is termed as shot crete and the lining is known as shot crete lining. The process is also known as gunning, as a gun is used for laying the mortar. Sometimes, this lining is known as gunned lining. The lining is done in two ways.

- By Dry Mix: In this method, a mixture of cement and moist sand is prepared and loaded in the cement gun. Then it is forced through the nozzle of the gun with the help of compressed air. The mortar spreads over the sub-grade to a thickness which varies from 2.5 cm to 5 cm.
- By Wet Mix: In this process, the mixture of cement, sand and water is prepared according to the approved consistency. The mixture is loaded in the gun and forced on the subgrade. This type of lining is very costly and it is not durable. It is suitable for resurfacing the old cement concrete lining.

8. Asphalt Lining

This lining is prepared by spraying asphalt (i.e. bitumen) at a very high temperature (about 150°C) on the subgrade to a thickness varies from 3 mm to 6 mm. The hot asphalt when becomes cold forms a water proof membrane over the subgrade. This membrane is covered with a layer of earth and gravel. The lining is very cheap and can control the seepage of water very effectively but it cannot control the growth of weeds.

9. Bentonite and Clay Lining

In this lining a mixture of bentonite and clay are mixed thoroughly to form a sticky mass. This mass is spread over the subgrade to form an impervious membrane which is effective in controlling the seepage of water, but it cannot control the growth of weeds. This lining is generally recommended for small channels.

10. Soil-Cement Lining

This lining is prepared with a mixture of soil and cement. The usual quantity of cement is 10 per cent of the weight of dry soil. The soil and cement are thoroughly mixed to get an uniform texture. The mixture is laid on the sub-grade and it is made thoroughly compact. The lining is efficient to control the seepage of water, but it cannot control the growth of weeds. So, this is recommended for small channels only.

1.9 Design of Lined Canals

The lined canals are not designed by the use of Lacey's and Kennedy's theory, because the section of the canal is rigid. Manning's equation is useful for designing.

The design considerations are:

- The section should be economical.
- The velocity should be maximum so that the cross-sectional area becomes minimum.
- The capacity of lined section is not reduced by silting.

Design Parameters for Circular Section: The bed is circular with its center at the full supply level and radius equal to full supply depth 'D'. The sides are tangential to the curve. However, the side slope is generally taken as 1:1.

Trapezoidal section: The horizontal bed is joined with side slope by a curve of radius equal to full supply depth D. The side slope is generally kept as 1:1.

Lined canals are built for five primary reasons:

- To permit the transmission of water at high velocities through areas of deep or difficult excavation in a cost - effective fashion.
- To permit the transmission of water at high velocity at a reduced construction cost.
- To decrease canal seepage, thus conserving water and reducing the waterlogging of lands adjacent to the canal.
- To reduce the annual cost of operation and maintenance.
- To ensure the stability of the channel section.

The design of lined channels from the view point of hydraulic engineering is a rather elementary process which generally consists of proportioning an assumed channel cross section. A recommended procedure for proportioning a lined section is summarized in table given below.

In this table, it is assumed that the design flow Q_D, the longitudinal slope of the channel S_0, the type of channel cross section e.g., trapezoidal and the lining material have all been selected prior to the initiation of the channel design process.

Step	Process
1	Estimate n or C for the specified lining material and S_0.
2	Compute the value of section factor $AR^{2/3} = nQ/\sqrt{S_o}$ or $AR^{1/2} = Q/\left(C\sqrt{S_o}\right)$.
3	Solve section factor equation for y_n given appropriate expressions for A and R.
4	If hydraulically efficient section is required, then the standard geometric characteristics are used and y_n is to be computed.

5	Check for, 1. Minimum permissible velocity if water carries silt and for vegetation. 2. Froude number (Check Froude number and other velocity constraints such as ≤ 0.8 V ≤ 5.5 m/s Generally, Froude number should be as small as possible for Irrigation canals and should be less than 0.35. Higher Froude numbers is permitted as in the case of super critical flows such as in chutes, flumes. Decide the dimensions based on practicability.
6	Estimate, 1. Required height of lining above water surface, 2. Required freeboard, Figure. Balance excavations costs, costs of channel lining and assess the needs to modify "Hydraulically efficient section".
7	Summarize the results with dimensioned sketch.

2

Water Logging and Drainage

2.1 Reclamation of Water Logged and Saline Soils: Causes and Control of Water Logging

Reclamation of Water Logged and Saline Soils

Water logged areas can be reclaimed by the following techniques:

- Proper Drainage System: The farmers should have adequate surface drainage facilities to remove excess water from their fields. The surface runoff and sub-soil drainage of the water should not be so slow. During rainy season efforts should be made not to retain water on soil surface.
- Using Tube Wells: A tube well is an ideal device to lower the level of water in water logged areas. They have the capability to draw out of the earth large quantities of water continuously. It is a good technique to reclaim water logged areas by installing the tube wells.
- Lining of Canals: In order to minimize water logging, concrete lining of canals and other water channels should be done. It will be helpful not only in controlling water logging but also in saving useful irrigation water.
- Water Management: Farmers should be educated about water management. Use of excessive irrigation water for cultivation of certain crops should be avoided. The modern irrigation techniques like drip irrigation should be adopted.
- Tolerant Crops: Crops like rice, oats, etc. should be preferred in water logged areas. Because the rice require more moisture for its growth.
- Tolerant Trees: Trees like Eucalyptus, willows, etc. should be planted in water logged areas because of its high moisture requirement.

Reclamation of Saline Soils

The amount of crop yield reduction depends on such factors as crop growth, the salt content of the soil, climatic conditions, etc. In extreme cases where the concentration of salts in the root zone is very high, crop growth may be entirely prevented. To improve crop growth in such soils the excess salts must be removed from the root zone.

The term reclamation of the saline soils refers to the methods used to remove soluble salts from the root zone. Methods commonly adopted or proposed to accomplish this include the following:

- Scraping: Removing the salts that have accumulated on the soil surface by mechanical means has only a limited success although many farmers have resorted to this procedure. Although this method might temporarily improve crop growth, the ultimate disposal of salts still poses a major problem.

- Flushing: Washing away the surface accumulated salts by flushing water over the surface is sometimes used to desalinize soils having surface salt crusts. Because the amount of salts that can be flushed from a soil is rather small, this method does not have much practical significance.

- Leaching: This is the most effective procedure for removing salts from the root zone of soils. Leaching is most often accomplished by ponding fresh water on the soil surface and allowing it to infiltrate.

It is effective when the salty drainage water is discharged through subsurface drains that carry the leached salts out of the area under reclamation. Leaching may reduce salinity levels in the absence of artificial drains when there is sufficient natural drainage, i.e. the ponded water drains without raising the water table.

Leaching should preferably be done when the soil moisture content is low and the groundwater table is deep. Leaching during the summer months is, as a rule, less effective because large quantities of water are lost by evaporation. The actual choice will although depend on the availability of water and other considerations.

In some parts of India for example, leaching is best accomplished during the summer months because this is the time when the water table is deepest and the soil is dry. This is also the only time when large quantities of fresh water can be diverted for reclamation purposes.

2.1.1 Causes and Control of Water Logging

Water Logging

In all surface water irrigation schemes, supplying the full water requirements of a crop, more water is added to the soil than is actually required to make up the deficit in the soil resulting from continuous evapotranspiration by crops. This excess water and the water that seeps into the ground from reservoirs, canals and watercourses percolate deep into the ground to join the water table and thus, raise the water table of the area. When the rising water table reaches the root zone, the pore spaces of the root-zone soil get saturated.

A land is said to be waterlogged when the pores of soil within the root zone of a plant gets saturated and the normal growth of the plant is adversely affected due to insufficient air circulation. The depth of the water table at which it starts affecting the plant would depend on plant and soil characteristics.

A land would become waterlogged sooner for deep-rooted plants than for shallow-rooted plants. Impermeable soils generally have higher capillary rise and hence, are waterlogged more easily than permeable soils. A land is generally waterlogged when the ground water table is within 1.5 to 2.0 m below the ground surface.

Water table depth is good if the water table is below 2 m and rises to 1.8 m for a period not exceeding 30 days in a year. If the water table is at about 1.8 m and rises to about 1.2 m for a period not exceeding 30 days in a year, the condition is considered as fair. If the water table depth is between 1.2 to 1.8 m which may rise to 0.9 m for a period not exceeding 30 days in a year, the condition of water table depth is rather poor.

In a poor condition of water table depth, the water level is less than 1.2 m from the surface and is generally rising. A high water table increases the moisture content of the unsaturated surface soil and thus increases the permeability.

There may be advantages of having water table close to the surface as it may result in higher crop yield due to favorable moisture supply. This may, however, be true only for few years after water table has risen from great depths.

The favorable condition may be followed by serious decrease in the crop yield in areas where alkali salts are present. With slight increase in inflow to the ground, the high water table may become too close to the ground surface and when this happens the land gets waterlogged and becomes unsuitable for cultivation.

The problem of water logging is a world-wide phenomenon which occurs mainly due to the rise of the ground water table beyond permissible limits on account of the change in ground water balance brought about by the percolation of irrigation water. It has become a problem of great importance on account of the introduction of big irrigation projects. The land subjected to waterlogging results in reduction of agricultural production. The problem of waterlogging has already affected about 5 million hectares of culturable area in India.

Causes of Waterlogging

Ground water reservoirs receive their supplies through percolation of water from the ground surface. This water may be from rainfall, from lakes or water applied to the fields for irrigation. This water percolates down to the water table and thus, raises its position.

Depending upon the elevation and the gradient of the water table, the flow may either be from surface to the ground (i.e., inflow) or ground to the surface (i.e., outflow). Outflow from a ground water reservoir includes water withdrawn through wells and water used as consumptive use. An overall balance between the inflow and outflow of a ground water reservoir will keep the water table at almost fixed level.

This balance is greatly disturbed by the introduction of a canal system or a well system for irrigation. While the former tends to raise the water table, the latter tends to lower it. Waterlogging in any particular area is the result of several contributing factors. The main causes of waterlogging can be grouped into two categories:

- Natural
- Artificial

Natural Causes of Waterlogging

Topography, geological features and rainfall characteristics of an area can be the natural causes of waterlogging. In steep terrain, the water is drained out quickly and hence, chances of waterlogging are relatively low. But in flat topography, the disposal of excess water is delayed and this water stands on the ground for a longer duration. This increases the percolation of water into the ground and the chances of waterlogging.

The geological features of subsoil have considerable influence on waterlogging. If the top layer of the soil is underlain by an impervious stratum, the tendency of the area getting waterlogged increases. Rainfall is the major contributing factor to the natural causes of waterlogging. Low lying basins receiving excessive rainfall have a tendency to retain water for a longer period of time and thus gets, waterlogged.

Submergence of lands during floods encourages the growth of weeds and marshy grasses which obstruct the drainage of water. This, again, increases the amount of percolation of water into the ground and the chances of waterlogging.

Artificial Causes of Waterlogging

There exists a natural balance between the inflow and outflow of a ground water reservoir. This balance is greatly disturbed due to the introduction of artificial irrigation facilities. The surface reservoir water and the canal water seeping into the ground increase the inflow to the ground water reservoir. This raises the water table and the area may become waterlogged. Besides, defective method of cultivation, defective irrigation practices and blocking of natural drainage further add to the problem of waterlogging.

Effects of Waterlogging

The crop yield is considerably reduced in a waterlogged area due to the following adverse effects of waterlogging:

- Absence of soil aeration.
- Difficulty in cultivation operations.
- Weed growth.
- Accumulation of salts.

In addition, the increased dampness of the waterlogged area adversely affects the health of the persons living in that area.

Absence of Soil Aeration

In waterlogged lands, the soil pores within the root zone of crops are saturated and air circulation is cut off. Waterlogging, therefore, prevents free circulation of air in the root zone. Thus, waterlogging adversely affects the chemical processes and the bacterial activities which are essential for the proper growth of a plant. As a result, the yield of the crop is reduced considerably.

Difficulty in Cultivation

For optimum results in crop production, the land has to be prepared. The preparation of land (i.e., carrying out operations such as tillage, etc.) in wet condition is difficult and expensive. As a result, cultivation may be delayed and the crop yield is adversely affected. The delayed arrival of the crop in the market brings less returns to the farmer.

Weed Growth

There are certain types of plants and grasses which grow rapidly in marshy lands. In waterlogged lands, these plants compete with the desired useful crop. Thus, the yield of the desired useful crop is adversely affected.

Accumulation of Salts

As a result of the high water table in waterlogged areas, there is an upward capillary flow of water to the land surface where water gets evaporated. The water moving upward brings with it soluble salts from salty soil layers well below the surface.

These soluble salts carried by the upward moving water are left behind in the root zone when this water evaporates. The accumulation of these salts in the root zone of the soil may affect the crop yield considerably.

Remedial Measures for Waterlogging

The main cause of waterlogging in an area is the introduction of canal irrigation there. It is, therefore, better to plan the irrigation scheme in such a way that the land is prevented from getting waterlogged.

Measures, such as controlling the intensity of irrigation, provision of intercepting drains, keeping the full supply level of channels as low as possible, encouraging economical use of water, removing obstructions in natural drainage, rotation of crops, running of canals by rotation, etc., can help considerably in preventing the area from getting waterlogged.

In areas where the water table is relatively high, canal irrigation schemes should be planned for relatively low intensity of irrigation. In such areas canal irrigation should be allowed in the Kharif season only. Rabi irrigation should be carried out using ground water. Intercepting drains provided along canals with high embankments collect the canal water seeping through the embankments and thus, prevent the seeping water from entering the ground.

The full supply level in the channels may be kept as low as possible to reduce the seepage losses. The level should, however, be high enough to permit flow irrigation for most of the command area of the channel. For every crop there is an optimum water requirement for the maximum yield. The farmers must be made aware that the excessive use of water would harm the crop rather than benefit it.

The leveling of farm land for irrigation and a more efficient irrigation system decrease percolation to the ground and reduce the chances of waterlogging. The improvement in the existing natural drainage would reduce the amount of surface water percolating into the ground. A judicious rotation of crops can also help in reducing the chances of waterlogging.

Running of canals by rotation means that the canals are run for few days and then kept dry for some days. This means that there would not be seepage for those days when the canal is dry. This, of course, is feasible only in case of distributaries and watercourses. The combined use of surface and subsurface water resources of a given area in a judicious manner to derive maximum benefits is called conjunctive use of water.

During dry periods, the use of ground water is increased and this results in lowering of the water table. The use of surface water is increased during the wet season. Because of the lowered water table, the ground water reservoir receives rainfall supplies through increased percolation. The utilization of water resources in this manner results neither in excessive lowering of the water table nor in its excessive rising.

The conjunctive use of surface and subsurface water serves as a precautionary measure against waterlogging. It helps in greater water conservation and lower evapotranspiration losses and brings larger areas under irrigation.

The most effective method of preventing waterlogging in a canal irrigated area, however, is to eliminate or reduce the seepage of canal water into the ground. This can be achieved by the lining of irrigation channels (including watercourses, if feasible). In areas which have already become waterlogged, curative methods such as surface and subsurface drainage and pumping of ground water are useful.

2.2 Reclamation of Saline and Alkaline Land

The excessive accumulation of alkali salts in the soils is injurious for plants growth It is therefore necessary, to reduce the percentage of salts to optimum or normal level so that plants may grow luxuriantly in such soils.

There are several methods of reclamation which can be grouped as follows:

- Chemical method in which some chemicals are added to the soil in order to bring the alkalinity to desired level.
- Mechanical practices such as improving drainage and leaching, mechanical shattering of clay pans and scrapping.
- Cultural method (growing salt tolerant plants).

Since fundamental causes in various groups of salty soils are different, their reclaiming techniques are different.

1. Reclamation of Alkali Soils

Alkali soils are best reclaimed by the following methods:

(i) Chemical Method

By cationic exchange (replacement of alkali from soil colloids by calcium ions). Application of calcium sulphate (gypsum) in the soil reduces alkalinity to a great extent and makes the soil fertile.

The reaction proceeds in the following way,

$$2\text{Na} - \text{clay} + \text{CaCo}_3 \rightarrow \text{Na}_2\text{SO}_4\text{Calcium} - \text{clay}$$

$$\text{Na}_2\text{CO}_3 + \text{CaSO}_4 \rightarrow \text{CaCO}_3 + \text{Na}_2\text{SO}_4.$$

Good drainage leaches away Na_2SO_4.

Alkali salt percentage can also be reduced in the soil by the use of acid forming chemical amendments such as sulphur, ferrous sulphate and limestone. Sulphur, when applied to the soil, oxidizes and forms sulphuric acid which converts carbonates of sodium and potassium to Na_2SO_4 and K_2SO_4 respectively that may be removed from top soil by drainage water.

The amount of gypsum and sulphur required to reclaim the alkali soils will be different depending upon the degree of alkalinity, drainage and buffering capacity of soils.

The types of reaction which occur when an amendment is applied to an alkali soil are given below:

With Sulphur:

- $2S + 3O_2 \rightarrow 2SO_3$
- $SO_3 + H_2O \rightarrow H_2SO_4$

In the next step, if soil is calcareous,

- $H_2SO_4 + CaCO_3 \rightarrow CaSO_4 + CO_2 + H_2O$
- $2Na - Clay + CaSO_4 \rightarrow Ca - Clay + Na_2SO_4$

But if the soil is non-calcareous,

- $2Na - Clay + H_2SO_4 \rightarrow 2H - clay + Na_2SO_4$

With lime-sulphur:

- $CaS_5 + O_2 + 4H_2O \rightarrow CaSO_4 + 4H_2SO_4$

Now if the soil is calcareous,

- $H_2SO_4 + CaCO_3 \rightarrow CaSO_4 + CO_2 + H_2O$
- $2Na - Clay + CaSO_4 \rightarrow Ca - Clay + Na_2SO_4$

But if the soil is non-calcareous,

- $10Na - Clay + 4H_2SO_4 + CaSO_4 \rightarrow 8H - Clay + Ca - Clay + 5\,Na_2SO_4$

With ferrous sulphates:

- $FeSO_4 + H_2O \rightarrow H_2SO_4 + FeO$

Now if the soil is calcareous,

- $H_2SO_4 + CaCO_3 \rightarrow CaSO_4 + CO_2 + H_2O$
- $2Na - Clay + CaSO_4 \rightarrow Ca - Clay + Na_2SO_4$

But if the soil is non-calcareous,

- $2Na-Clay+H_2SO_4 \rightarrow 2H-Clay+Na_2SO_4$

With limestone on non-calcareous soils:

- $Na-Clay+H_2O \rightarrow H-Clay+NaOH$
- $2H-Clay+2NaOH+CaCO_3 \rightarrow Ca-Clay+Na_2CO_3+2H_2O$

With any H-Clay:

- $2H-Clay+CaCO_3 \rightarrow Ca-Clay+CO_2+H_2O$

The use of pyrite (FeS_2) as an amendment is a recent development in the chemical amelioration and reclamation of alkali soils. In presence of moisture and air, pyrite is converted into sulphuric acid which then replaces exchangeable sodium by hydrogen or calcium released from insoluble calcium present in the soil.

In addition it is said to correct iron deficiency and lime induced iron chlorosis in alkali soils. It is important to mention that the formation of H_2SO_4 in the soil by the application of pyrite may take place through chemical and microbiological actions. Pyrite is oxidized according to the following equation suggested by Bloomfield (1973),

$$FeS_2+2Fe+3 \rightarrow 3Fe+2+2s \text{ (Chemical)}$$

Sulphur, thus formed could be the substrate for thioxidants which convert it into H_2SO_4,

$$S+3(O)+H_2O \rightarrow H_2SO_4$$

Temple and Kochler (1954) explained the action of ferroxidans on the formation of H_2SO_4 as follows,

$$FeS_2+H_2O+7O \rightarrow FeSO4+H_2SO_4$$

$$2FeSO_4+O+H_2SO_4 \rightarrow Fe_2(SO_4)3+H_2O$$

$FeSO_4$ formed in the above reaction may be converted into H_2SO_4 by hydrolysis,

$$FeSO_4+H_2O \rightarrow H_2SO_4+FeO$$

The pyrite is oxidized in soils to ferrous sulphate and sulphuric acid as depicted in the following equation,

$$2FeS_2+2H_2O+7O_2 \rightarrow 2FeSO_4+2H_2SO_4$$

Both sulphuric acid and ferrous sulphate help in reclamation of calcareous as well as non- calcareous salt affected soils by lowering the pH and solubilizing free calcium from calcium carbonate present.

The reactions are given below:

In salt affected calcareous soils,

$$CaCO_3 + H_2SO_4 \rightarrow CaSO_4 + CO_2 + H_2O$$

$$2Na-Clay + CaSO_4 \rightarrow Ca-Clay + Na_2SO_4$$

$$FeSO_4 + H_2O \rightarrow FeO + H_2SO_4$$

H_2SO_4 formed in reaction II reacts as per equations la and lb

H_2SO_4 also neutralizes $NaHCO_3$ and Na_2CO_3 present in these soils,

$$2NaHCO_3 + H_2SO_4 \rightarrow Na_2SO_4 + 2H_2O + 2CO_2$$

$$Na_2CO_3 + H_2SO_4 \rightarrow Na_2SO_4 + H_2O + CO_2$$

But if the soil is non-calcareous:

$$2Na-Clay + H_2SO_4 \rightarrow 2H-Clay + Na_2SO_4$$

$$FeSO_4 + H_2O \rightarrow FeO + H_2SO_4$$

H_2SO_4 formed in reaction II acts in similar manner as in reaction I.

Dhar's method: In India, Dr. Neel Ratan Dhar (1935) succeeded in reducing the alkalinity and salinity of the soil by the use of molasses and press-mud.

For one acre land he recommended the mixture of the following substances:

- 2 tons of molasses.
- 1-2 tons of press-mud (a waste product of sugar industry).
- 50-100 pounds of P_2O_5 in the form of basic slag.

The molasses is fermented by soil microbes and as a result of fermentation organic acids are produced which lower the alkalinity and increase the availability of phosphates. The press-mud contains Ca which forms calcium salts that reduce the content of exchangeable sodium. Phosphate helps in the microbial fixation of nitrogen into nitrogenous compounds in the soil.

(ii) Mechanical Methods

The alkali salts are removed by:

- Scraper or by rapidly moving streams of water.
- Deep ploughing of the land which reduces the alkalinity and makes the soil more permeable.

- Application of green manures of Dhaincha, guar, jantar (Sesbania aculeata) has been found most successful in reclamation of alkali and saline soils.
- Spreading of straw and dried grasses and leaves on the alkaline soil.

(iii) Cultural Method

Growing of alkali tolerant crops and plants, such as sugar-beet, rice, persian (Hibiscus cannabinus), wild indigo and babul in such soils successfully reduces the alkalinity. The rice is commonly the first crop grown on salty lands to be reclaimed.

In Punjab the usual practice of reclamation of salty lands involves growing of paddy after first initial leaching followed by berseem or senji which has higher water requirement than Dhaincha as green manure which is followed by sugarcane and then wheat or cotton.

Introduction of leguminous crops helps in building up of nitrogen supply and opens the soils. Dhaincha-paddy-berseem rotation has been found to be the best cropping pattern on mild type of alkali soils in Punjab region.

In U.P. also, paddy or dhaincha-paddy are the usual crops taken during first stage of reclamation of salty soils. This is followed by berseem or barley in winter. Pulse crops like gram or peas show poor performance.

2. Reclamation of Saline Soil

Saline soil can be reclaimed by the following methods:

- By lowering the water table 5-6 feet below the surface. In sloppy area, it can be done by making network of 5-6 feet deep trenches at right angles to the slopes. In course of 2 or 3 Successive leaching, harmful salts are removed. A deep ploughing is also helpful in reclamation of saline soil. This also makes the soil loose and thus facilitates the downward movement of salty water in the soil.
- Salt tolerant crops, e.g., rice, sugarcane, barley and castor gradually remove salts from the soil.
- In case of saline soils which do not contain calcium salts, the addition of $CaSO_4$ (gypsum) is beneficial. Supply of calcium in the soil can indirectly be maintained by organic matters which on decomposition produce CO_2. The CO_2 gas, so produced, combines with insoluble calcium carbonate in moist condition to form soluble calcium bicarbonate. This also reduces alkalinity.
- Application of green manure, organic manures, organic residues, acids or acid formers is yet another good way to reduce salinity.

3. Reclamation of Saline-Alkaline Soil

Here the problem of reclamation is two-fold because of:

- Heavy accumulation of different types of salts.
- Poor percolation due to the presence of hard clay pan and highly dispersed sodium clay.

Such soil can be reclaimed by:

- Mechanical shattering of clay pans. This helps in downward movement of water.
- Application of gypsum in the soil. This is followed by flushing with plenty of water.
- Green maturing with Dhaincha (Sesbenia aculeata).
- Growing of salt tolerant plants, e.g., paddy in kharif and oat and barley in rabi seasons are recommended for such soils.

Schoonover (1959) worked on the soils of India and enlisted the following technical requirements for reclamation of saline and alkaline soils:

- Necessity of good drainage.
- Availability of sufficient water to wash the excess salts from the top soils.
- Good soil management including land leveling, good bonding for irrigation and advanced agronomic practices.
- Protection of soil from erosions.
- Good quality of irrigation water.

2.2.2 Surface and Sub-Surface Drainage

Drainage

Drainage is defined as the removal of excess water and salts from adequately irrigated agricultural lands. The deep percolation losses from properly irrigated lands and seepage from reservoirs, canals and watercourses make drainage necessary to maintain soil productivity. Irrigation and drainage are complementary to each other.

In humid areas, drainage attains much greater importance than in arid regions. Irrigated lands require adequate drainage to remain capable of producing crops. The adequate drainage of fertile lands requires the lowering of a shallow water table and this forms the first and basic step in the reclamation of waterlogged, saline and alkali soils. The drainage of farm lands:

- Improves soil structure and increases the soil productivity.
- Facilitates early ploughing and planting.
- Increases the depth of root zone thereby increasing the available soil moisture and plant food.
- Increases soil ventilation.
- Increases water infiltration into the ground thereby decreasing soil erosion on the surface.
- Creates favourable conditions for growth of soil bacteria.
- Leaches excess salts from soil.
- Maintains favourable soil temperature.
- Improves sanitary and health conditions for the residents of the area.

The water table can be lowered by eliminating or controlling sources of excess water. An improvement in the natural drainage system and the provision of an artificial drainage system are of considerable help in the lowering of the water table. A natural drainage system can be properly maintained at low costs and is a feasible method of protecting irrigated lands from excessive percolation. Artificial drainage also aims at lowering the water table and is accomplished by any of the following methods:

- Open ditch drains or Surface drains.
- Subsurface drains.
- Drainage wells.

Open Ditch Drainage (or Open Drainage or Surface Drainage)

They are suitable and very often economical for surface and subsurface drainage. They permit easy entry of surface flow into the drains. Open drains are used to convey excess water to distant outlets. These accelerate the removal of storm water and thus reduce the detention time thereby decreasing the percolation of water into the ground. Open drains can be either shallow surface drains or deep open drains.

Shallow surface drains do not affect subsurface drainage. Deep open drains act as outlet drains for a closed drain system and collect surface drainage too. The alignment of open drains follows the paths of natural drainage and low contours. The drains are not aligned across a pond or marshy land. Every drain has an outlet, the elevation of which decides the bed and water surface elevations of the drain at maximum flow.

The longitudinal slope of drain should be as large as possible and is decided on the basis of non-scouring velocities. The bed slope ranges from 0.0005 to 0.0015. Depths of about 1.5 to 3.5 m are generally adopted for open drains. The side slopes depend largely on the type of embankment soil and may vary from 1/2 H: 1V (in very stiff and compact clays) to 3H: 1V (in loose sandy formations).

The cross-section of open drain is decided using the general principles of channel design. The channel will be in cutting and the height of banks will be small. If the drain has to receive both seepage and storm water, it may be desirable to have a small drain in the bed of a large open drain. This will keep the bed of the drain dry for most of the year and maintenance problems will be considerably less. Only the central deeper section will require maintenance. Open drains have the advantages of:

- Low initial cost.
- Simple construction.
- Large capacity to handle surface runoff caused by precipitation.

However, there are disadvantages too. Besides the cost of land which the open drains occupy and the need of constructing bridges across them, open drains cause:

- Difficulty in farming operations.
- Constant maintenance problems resulting from silt accumulation due to rapid weed growth in them.

Flow of clear water at low velocities permits considerable weed growth on the channel surface. The open drains have, therefore, to be cleaned frequently. In addition to manual cleaning, chemical weed killers are also used. But, at times the drain water is being used for cattle and the weed poison may be harmful to the cattle. Aquatic life is also adversely affected by the chemical weed killers.

Subsurface Drainage (or Under Drainage)

It involves the creation of permanent drainage system consisting of buried pipes (or channels) which remain out of sight and therefore, do not interfere with the farming operations. The buried drainage system can remove excess water without occupying the land area. Therefore, there is no loss of farming area. Besides, there is no weed growth and no accumulation of rubbish and therefore, the under drainage system can remain effective for long periods with little or no need for maintenance. In some situations, however, siltation and blockage may require costly and troublesome maintenance or even complete replacement.

The materials of the buried pipes include clay pipes and concrete pipes in short lengths (permitting water entry at the joints) or long perforated and flexible plastic pipes. In addition, blankets of gravel laid in the soil, fibrous wood materials buried in the soil or such materials which can be covered by the soil and which will remain porous for long time are used for the construction of under drainage system.

If such drains are to be placed in impervious soil, the drains should be surrounded by a filter of coarser material to increase the permeability and prevent migration of soil particles and blocking of drains. Mole drains are also included as subsurface drains.

The mole drains are unlined and unprotected channels of circular cross-section constructed in the subsoil at a depth of about 0.70 m by pulling a mole plough through the soil without digging a trench.

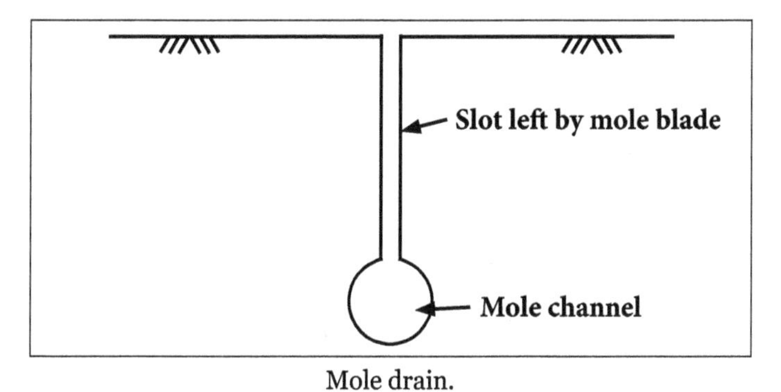

Mole drain.

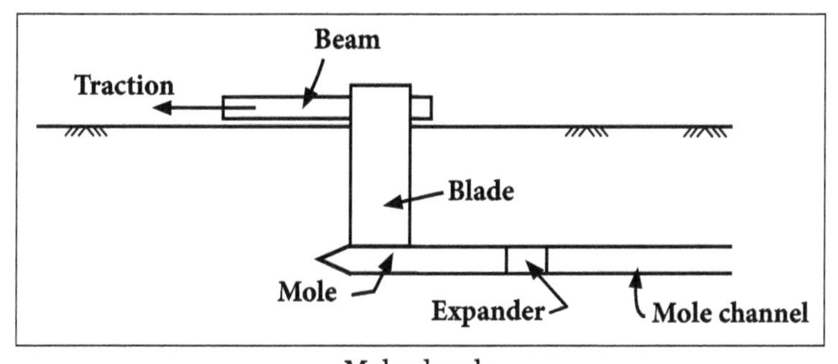

Mole plough.

The mole plough is a cylindrical metal object (about 300 to 650 mm long and 50 to 80 mm in diameter) with one of its ends bullet-shaped. The mole is attached to a horizontal beam through a thin blade as shown in Figure.

A short cylindrical metal core or sphere is attached to the rear of the mole by means of a chain. This expander helps in giving a smooth finish to the channel surface. The basic purpose of all these subsurface drains is to collect the water that flows in the subsurface region and to carry this water into an outlet channel or conveyance structure. The outlets can be either gravity outlets or pump outlets. The depth and spacing of the subsurface drains (and also deep open drains) are usually decided using Hooghoudt's equation described in the following.

Consider two drains at a spacing of B and the resulting drained water table as shown in figure below. An impermeable layer underlies the drain at a depth d. Rainfall intensity (or rate of application of irrigation water) is uniform and is equal to ra (m/s). Hooghoudt made the following assumptions to obtain a solution of the problem:

(i) The soil is homogeneous and isotropic.

(ii) The hydraulic gradient at any point is equal to the slope of the water table above the point, i.e., dh/dx.

(iii) Darcy's law is valid.

Using Darcy's law one can write,

$$qx = ky\frac{dy}{dx} \qquad ...(1)$$

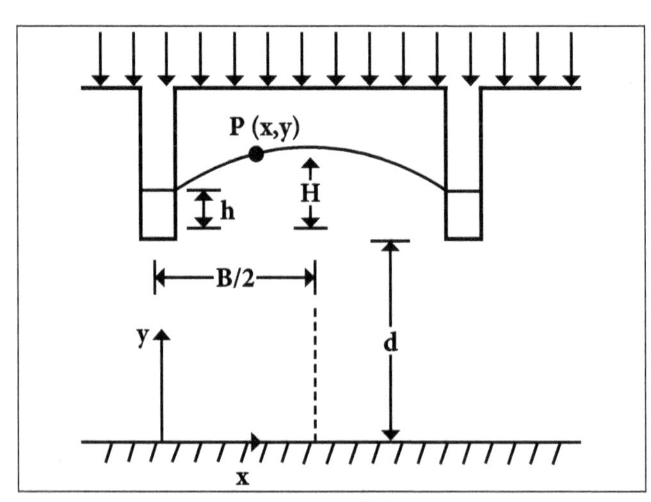

Deep open drains.

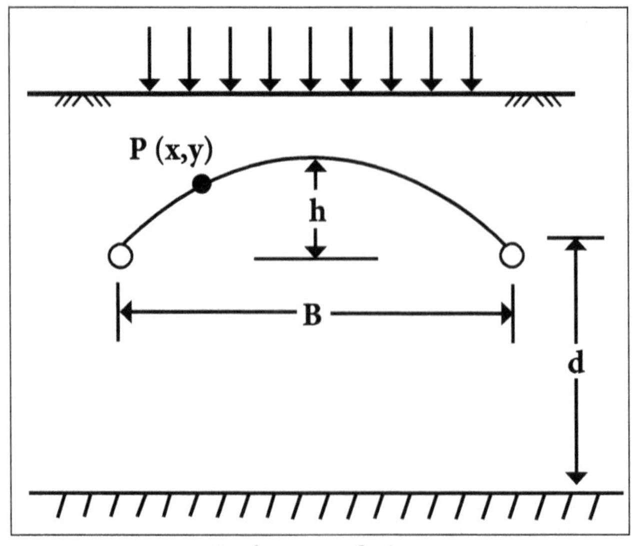

Surface open drains.

Where,

q_x - The discharge per unit length of drain at a section.

x - Distance away from the drain.

k - The coefficient of permeability of the soil.

$$q_x = \left(\frac{B}{2} - x\right) r_a \quad ...(1)$$

Using Equations (1) and (2), one can write,

$$\left(\frac{B}{2} - x\right) r_a = ky\frac{dy}{dx}$$

$$\frac{B}{2} r_a dx - r_a x dx = ky\ dy$$

On integrating,

$$r_a \frac{B}{2} x - r_a \frac{x^2}{2} = k\frac{y^2}{2} + C$$

The constant of integration C can be determined by using the boundary condition:

at x = 0, y = h + d

$$C = -\frac{k(h+d)^2}{2}$$

$$r_a \frac{B}{2} x - r_a \frac{x^2}{2} = \frac{k}{2}\left[y^2 - (h+d)^2\right]$$

Further,

at x=B/2, y = H + d

$$r_a \frac{B^2}{4} - r_a \frac{B^2}{8} = \frac{k}{2}\left[(H+d)^2 - (h+d^2)\right]$$

or,

$$r_a \frac{B^2}{4} - k\left[(H+d)^2 - (h+d^2)\right]$$

$$B^2 = \frac{4k}{r}\left[(H+d)^2 - (h+d)^2\right] \quad ...(3)$$

Equation (3) is Hooghoudt's equation for either open ditch drains or subsurface drains.

If q_d is the discharge per unit length of drain that enters the drain from two sides of the drain.

Then,

$$q_d = r_a B$$

$$q_d = \frac{4k}{B}\left[(H+d)^2 - (h+d)^2\right] \qquad ...(4)$$

In practice, the drain is considered empty (i.e., h = 0). Equation (4) then reduces to,

$$q_d = r_a B$$

$$q_d = \frac{4k}{B}\left[(H+d)^2 - (h+d)^2\right] \qquad ...(5)$$

$$q_d = \frac{4k\left[(H+d)^2 - (d)^2\right]}{B} \qquad ...(6)$$

$$q_d = \frac{4kH}{B}(H+2d)$$

or,

$$B^2 = \frac{4kH}{r_a}(H+2d) \qquad ...(7)$$

Thus, knowing q_d, one can determine the spacing B of the drains. The design drain discharge (or the drainage coefficient which is defined as the amount of water that must be removed in a 24-hour period) primarily depends on the rainfall rate, size of the watershed and the amount of surface drainage water that is admitted to the drainage system and is usually taken as equal to one percent of average rainfall in one day.

Thus,

$$q_d = \frac{0.01 \times r_a \times B}{24 \times 3600} m^{3/s}$$

Here, r_a is the average rainfall intensity in meters and B is the spacing of the drains in meters. The value of B is generally between 15 and 45 m. The main drawback of the gravity drainage system is that it is not capable of lowering the water table to large depths.

Drainage wells offer a very effective method of draining an irrigated land. The soil permeability and economic considerations decide the feasibility of well drainage. Drainage wells pump water from wells drilled or already existing in the area to be drained.

The above-mentioned remedial methods can be grouped as structural measures. In addition, the following non-structural measures can also be resorted to for preventing or reducing the menace of waterlogging:

- Adoption of tolerant crops.
- Restricting canal supplies close to crop-water needs.
- Switch over to drip irrigation.
- Conjunctive use of surface and ground water.
- Rationalization of water and power pricing policies.
- Improvement in canal irrigation management.
- Incentives for reclamation of land.

In arid regions, the bio-drainage (plantation of trees having high transpiration rates) would help in controlling the rise of ground water table and soil salinity. In addition, the bio-mass so grown acts as shelter belt in light soil area against shifting sands and dunes such as in Indira Gandhi Nahar Pariyojna (IGNP) command area in which eucalyptus trees and other trees of similar species were planted. The plantation was very effective in lowering of water table.

2.3 Types of Cross Drainage Works

It is a structure for carrying the discharge from a natural stream across the canal intercepting the stream. Canal comes across obstructions like natural drains, rivers and other canals. Several various types of structures that are built to carry canal water across the above mentioned obstructions or vice versa are known as cross drainage works.

There are many different factors that are involved in selecting a specific type of Cross drainage works and in selecting a suitable site for cross drainage works. It is generally a costly item and must be avoided by diverting one stream into another stream. Changing the alignment of canal so that it crosses below the junction of two streams.

Depending upon the levels and the discharge, types of cross drainage works can be of the following types:

Cross drainage works carrying a canal across the drainage.

Structures that fall under this type are as follows:

- An Aqueduct.
- Siphon Aqueduct.

Aqueduct

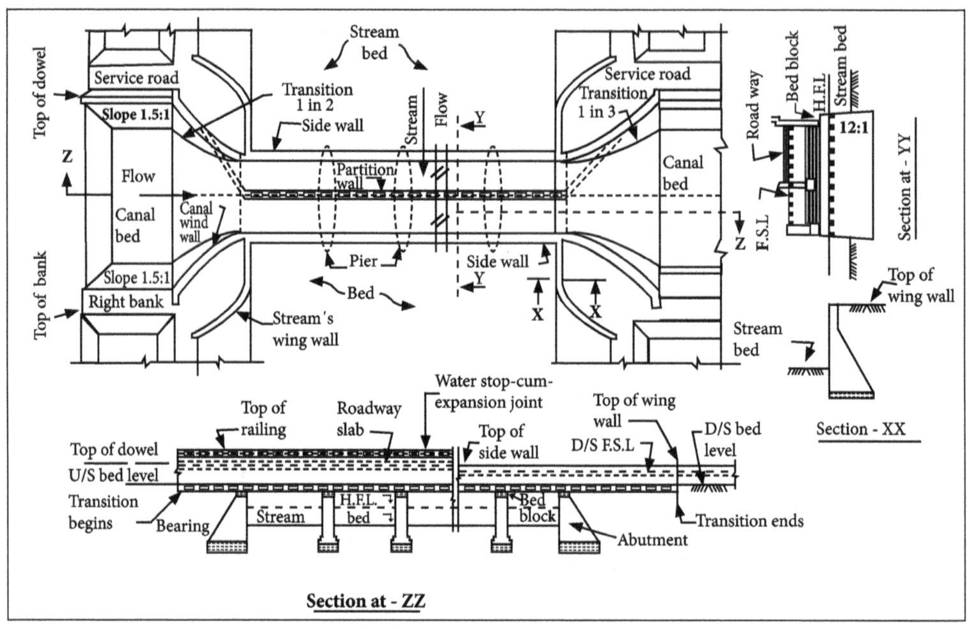

Typical section of an aqueduct.

When HFL of the drain is sufficiently below the bottom of canal in a manner that the drainage water flows freely under gravity, the structure is known as Aqueduct. In this, canal water is being carried across the drainage in a trough supported on the piers. Bridge carrying water provided when necessary level difference is available between the canal and canal bed is sufficiently higher than HFL.

Crossing works (Aqueduct).

Siphon Aqueduct

In case of the siphon Aqueduct, the HFL of drain is higher above the canal bed and water runs under the siphonic action through the aqueduct barrels. The drain bed is generally depressed and provided with pucca floors on the upstream side. Drainage bed may be joined to the pucca floor either by a vertical drop or by glacis of 3:1. The downstream rising slope must not be steeper than 5:1.

When canal is passed over the drain, the canal remains open for inspection throughout and damage caused by flood is rare. During heavy floods, the foundations are susceptible to scour or waterway of drain may get choked due to debris, tress etc.

Siphon Aqueduct.

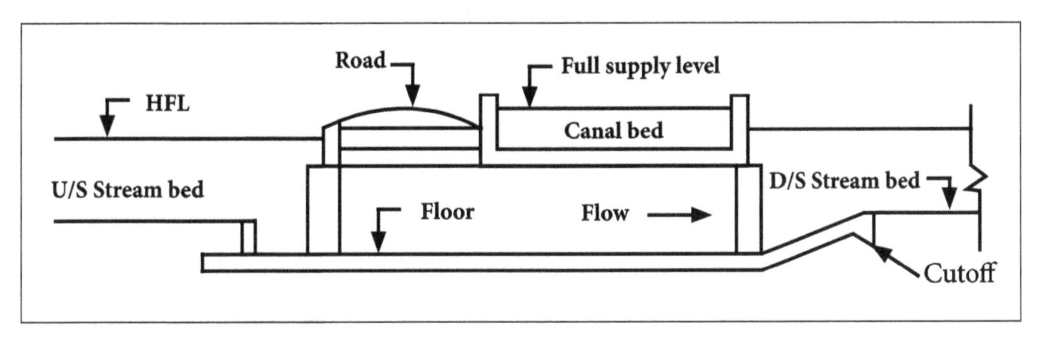

Layout of siphon aqueduct.

Type I: In this type of structure, earthen canal banks are carried as such and therefore, the culvert length has to be long enough to support the water section as well as the earthen banks of the canal. In this type of structure, canal section is not flumed and thus remains unaltered. Therefore, the width of the structure is maximum.

This type of structure, obviously, saves on the canal wings and bank connections and is justified only for small streams so that the length of the structure is small. An extreme example of this kind of structure would be when the stream is carried by means of a pipe laid under the bed of the canal.

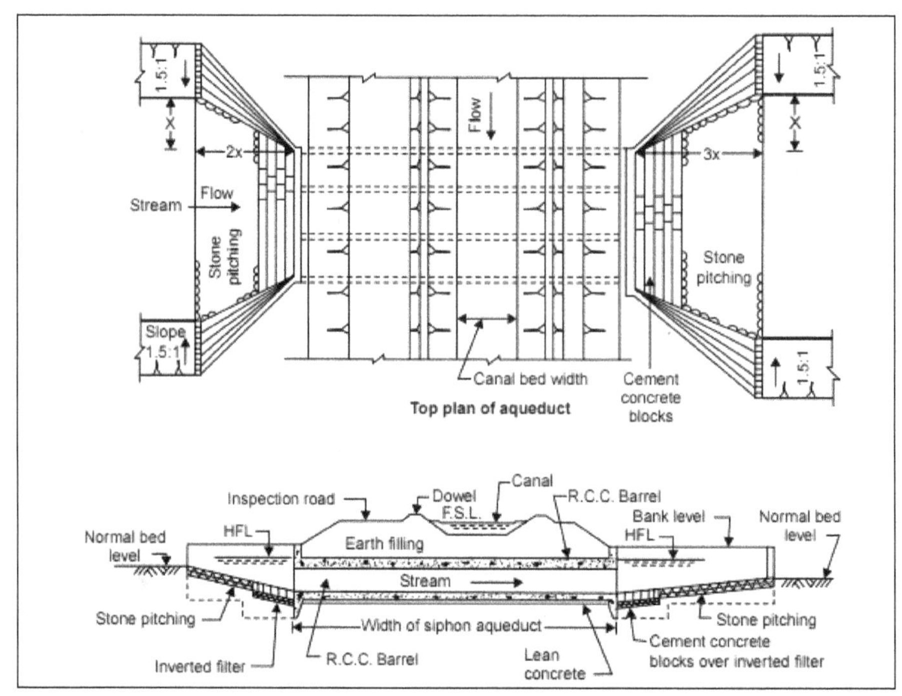

Typical section of siphon aqueduct (Type I).

Type II: This type of structure is similar to the Type I with a provision of retaining the walls to retain the outer slopes of earthen canal banks. This reduces the length of culvert. This type of construction may be considered suitable for the streams of intermediate sizes.

Type III: In this type of structure, the earthen canal banks are discontinued through aqueduct and the canal water is carried in a trough which may be of either masonry or concrete. The earthen canal banks are connected to the corresponding through walls on their sides by means of wing walls. The width of canal is also reduced over crossing.

In this type of structure, width of the structure is minimum and thus, the structure is suitable for large streams requiring considerable length of aqueduct between the abutments.

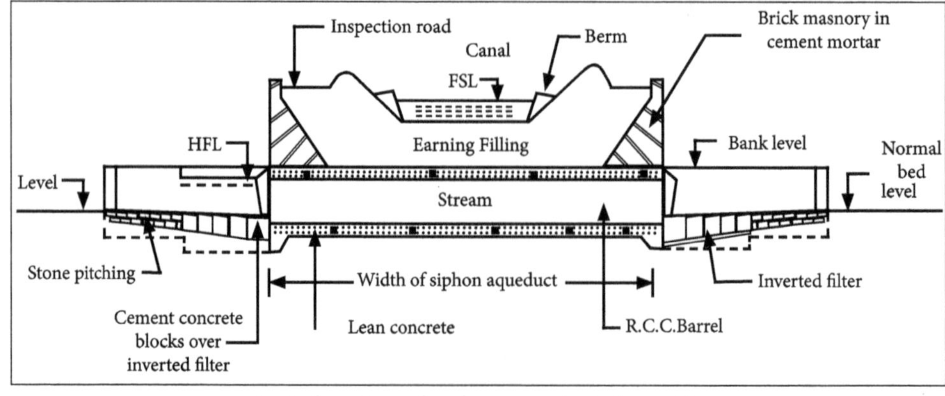

Typical section of siphon aqueduct (Type II).

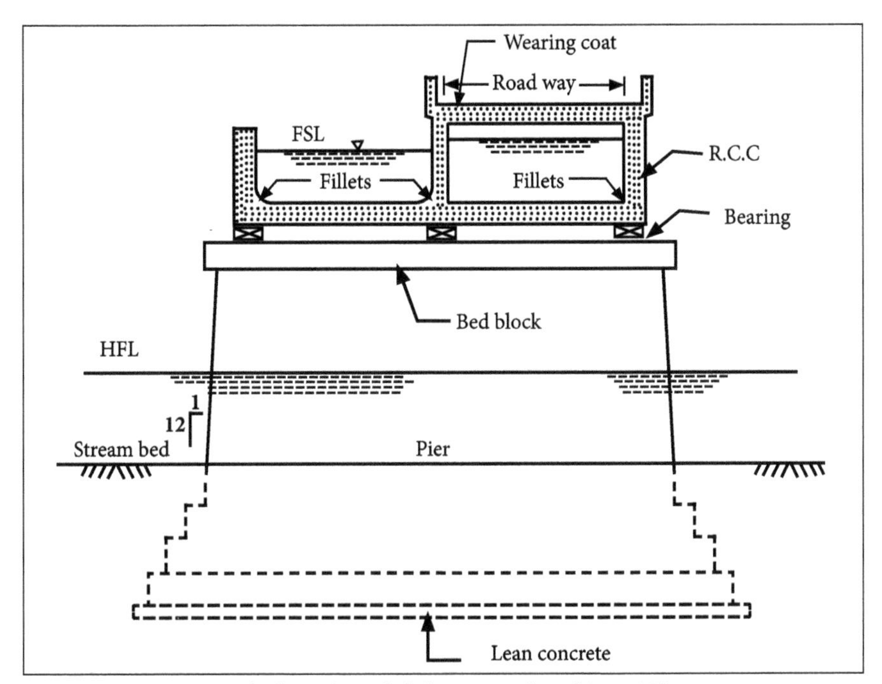

Typical section of siphon aqueduct (Type III).

Selection of Suitable Site for Cross Drainage Works

The factors that affect the selection of suitable type of cross drainage works are:

- Relative bed levels and water levels of canal and the drainage.
- Size of the canal and drainage.

The following considerations are important:

- When the bed level of the canal is well above the HFL of the drainage, an aqueduct is the obvious choice.
- When the bed level of the drain is well above the FSL of canal, super passage is provided.
- The necessary headway between canal bed level and the drainage HFL can be increased by shifting the crossing to the downstream of drainage. If, it is not possible to change canal alignment, a siphon aqueduct can be provided.
- When the canal bed level is much lower, but the FSL of canal is present higher than the bed level of drainage, a canal siphon is preferred.
- When the drainage and canal cross each other practically at the same level, a level crossing can be preferred. This type of work is avoided as far as possible.

Factors Which Influence the Choice/Selection of Cross Drainage Works

The considerations which govern the choice between the aqueduct and the siphon aqueduct are:

- Suitable canal alignment.
- Suitable soil available for bank connections.
- Nature of available foundations.
- Permissible head loss in canal.
- Availability of funds.

Compared to an aqueduct, a super passage is inferior and must be avoided whenever possible. Siphon aqueduct is preferred over siphon unless large drop in drainage bed is needed.

2.4 Design Consideration for CD Works

Design principles of aqueduct, siphon aqueduct and super passage:

1. Estimation of Design Flood Discharge of a Drain

The drain to be crossed can be small or like a river. In all cases correct assessment of maximum flood or peak flow of the drain must be obtained beforehand.

2. Waterway Requirement for a Drain

Lacey's regime perimeter equation holds good basis for calculating drainage waterway. The equation is given by,

$$P_w = 4.825\, Q^{1/2}$$

Where,

P_w - Waterway to be provided for drain at the site in meters.

Q - Flood discharge of drain in m^3/sec.

As piers reduce the actual waterway available, the length between abutments (P_w) can be increased by 20 percent. When the waterway is fixed from the Lacey's regime perimeter equation, regime condition in the drain upstream and the downstream of the structure is not disturbed appreciably. To confine drainage water to the desired waterway guide banks can be constructed.

3. Velocity of Flow Through Barrel

The velocity of flow through the barrel may range from 1.8 m/sec to 3 m/sec, the reason for selection of this range is that the lower velocities can cause silting in the barrels, Whereas when the velocity is higher than 3 m/sec the bed load can cause abrasion of the barrel floor and subsequently it can be damaged.

4. Height of Opening

Once the waterway discharge and velocity are fixed the depth of flow will be obtained easily. There must be sufficient headway or clearance left between the HFL and the bottom of canal bed. A clearance of 1 m or half the height of the culvert, whichever is less will be sufficient.

Thus,

Height of opening = Depth of flow + Clearance or headway.

5. Number of Spans

After determining the total length of an aqueduct between the abutments number of spans to be provided may be fixed on the basis of the following two considerations:

- Structural strength required.
- Economical consideration.

For example, when the arches are used the number of spans to be given can be more. When the cost of construction in the foundation is high, small number of spans is adopted and then RCC beams can be used.

6. Canal Waterway

Generally fluming ratio is taken to be 1/2. This ratio is adopted in such a way that the velocity of flow in trough does not go above critical velocity limit. Generally velocity of flow must not be more than 3 m/sec. This precaution is taken to avoid the possibility of formation of hydraulic jump. The obvious reason is that when hydraulic jump is formed it will absorb energy. In this process valuable head is lost and large stresses are produced in the structure.

7. Length of the Contraction or Approach Transition

Once the width at throat is fixed, length of contraction may be determined after knowing convergence ratio. The convergence ratio is generally taken as 2:1, i.e., not steeper than 30°.

8. Length of the Expansion or Departure Transition

Length of the expansion on the downstream side of the aqueduct can be fixed after knowing the expansion ratio. The expansion ratio is generally taken as 3:1, i.e., not steeper than 22.5°. To maintain the streamlined flow and also to reduce loss of head the transitions are generally made up of the curved and flared wing walls.

The design of the transition can be worked out by making use of any of the following three methods:

- Hind's method.
- Mitra's hyperbolic transition method.
- Chaturvedi's semi-cubical parabolic transition method.

It can be noted that Hind's method may be used when the water depth in the normal section and the flumed trough also varies, the remaining two methods may be used only when the water depth remains to be constant in the normal canal section as well as the trough section.

9. Bank Connections

An aqueduct requires four sets of wing wall, Canal wing walls on the upstream and downstream side of aqueduct protect and retain the earth in canal banks. The foundation of canal wing walls must not be left in the embanked earth. The wing walls must be based on the sound foundation in the natural ground. In the transitions the side slopes of the natural section are warped to conform with shape of the trough over the drain.

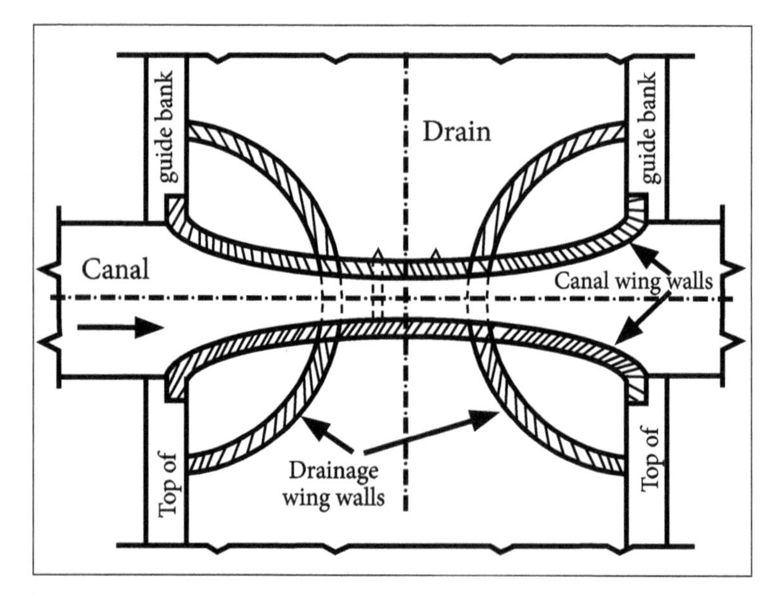

Drainage wing walls are provided on the upstream and downstream of the barrel to protect and to retain the natural sides of the drain. As the bed of the drain gets scoured during floods the drainage wing walls need to be taken deep into the foundation below maximum scour depth.

The wing walls should be taken back essentially into the top of guide banks. The wing walls must be designed to permit the smooth entry and exit of the flow in the drain.

Hind's Method for Design of Transition

This method is based on the premise that there is a minimum loss of head, flow is streamlined and normal flow conditions in the canal are restored before the canal discharges pass down on to the earthen section immediately after curved and in flared transitions.

In the contraction or approach transition, the throat portion and the expansion or departure transition are shown. It can be seen that sections 1-1, 2-2, 3-3 and 4-4 indicate the start of contraction, end of contraction, start of expansion and end of expansion respectively.

Thus the contraction or approach transition lies between sections 1 and 2, throat between the sections 2 and 3 and expansion or departure transition between sections 3 and 4. Up to section 1 and beyond section 4 canal flows under its normal conditions and therefore the canal parameters at these two points are equal and already known. So conditions of flow and canal parameters are same between sections 2 to 3 which represents throat or trough portion.

The design procedure can be outlined as follows:

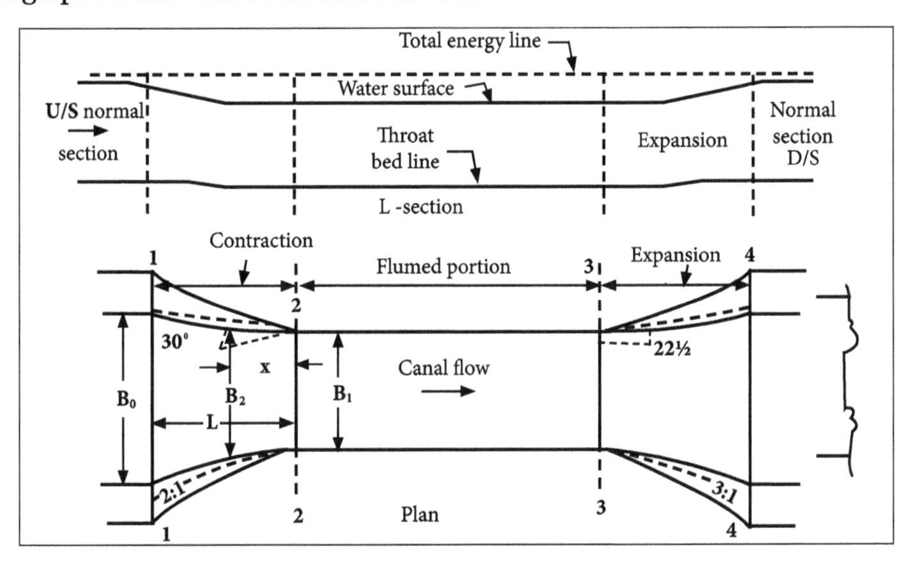

Let D and F with appropriate subscripts refers to depths and velocities at four sections also since canal levels and dimensions are already known in the section 4-4:

Step 1: TEL at section 4-4 = Elevation of Water surface V / 2g where water surface elevation at sec. 4-4 = Bed level + D_4

Step 2: TEL at section 3-3 = (TEL at section 4-4) + (energy loss between section 3 and 4) Energy loss between the sections 3-3 and 4-4 takes place due to the expansion of streamlines and also due to friction. Neglecting loss due to the friction which is small and taking loss due to the expansion to be.

$$0.3\left(\frac{V_3^2 - V_4^2}{2g}\right)$$

TEL at section 3-3=(TEL at section 4-4) + $0.3\left(\frac{V_3^2 - V_4^2}{2g}\right)$

Therefore, the Water surface level at section 3.3 = (TEL at section 4-4) - $\left(\frac{V_3^2}{2g}\right)$

And the bed level at section 3-3 = (Water surface level at section 3-3) - (D_3).

Step 3: Similarly, TEL at section 2-2 = (TEL at section 3-3) + (Head loss between the sections 2 and 3) Between sections 2-2 and 3-3 canal flows through trough section of same dimension and therefore flow is now uniform. Loss of head in the trough is only due to the friction losses. The loss of head can be calculated using Manning's equation.

$$Q = A.\frac{1}{N}R^{2/3}.S^{1/2}$$

Water surface level at section 2-2 = (TEL at section 2-2) - $\left(\frac{V_2^2}{2g}\right)$

Bed level at section 2-2 = (Water surface level at section 2.2) - (D_2)

Since depth and velocity are same in the trough from sections 2 to 3 it may be noted that the total energy line, water surface line and bed line are parallel to each other.

Step 4: On the similar lines,

TEL at section 1.1 = (TEL at section 2-2) + (Head loss between sections 1 and 2).

Between sections 2.2 and 1.1 energy loss takes place due to contraction of streamlines and friction. Neglecting friction loss, loss due to contraction may be taken to be $0.2\left(\frac{V_2^2 - V_1^2}{2g}\right)$.

TEL at section 1-1 = (TEL at section 2-2) + $0.2\left(\frac{V_2^2 - V_1^2}{2g}\right)$

Water surface level at section 1-1 = (TEL at section 1-1) - $\left(\frac{V_1^2}{2g}\right)$ and 2 bed level at section 1-1 = (Water surface level at section 1 1)-(D_1).

Step 5: As mentioned in the first four steps the bed level, water surface level and level of total energy line can be determined at the four sections.

The TE line, water surface line and the bed line can be drawn as follows:

- Now total energy line can be drawn by joining these points at four sections by a straight line.
- The bed line can also be drawn as straight lines between adjacent sections if the fall or rise of bed level is small. The corners must be rounded off. In case the drop in the bed line is appreciable the bed lines must be joined with a smooth tangential reverse curve.
- It is now clear that between any two consecutive sections the drop in water surface level may result due to:
    - Drop in the TE line between the two sections.
    - Increased velocity head in the contraction.
    - Decreased velocity head in the expansion.

This drop in water surface gets negotiated by two parabolic curves. As shown in Figures below for contraction and expansion, this is achieved by convex upwards curve followed by concave upwards curve in former transition and concave upwards curve followed by the convex upwards curve in the latter transition.

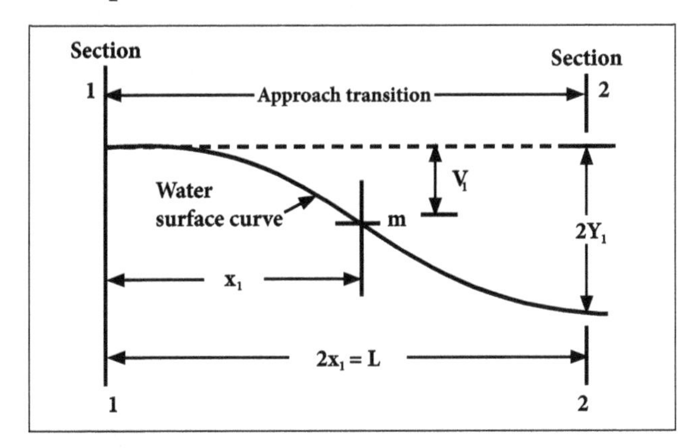

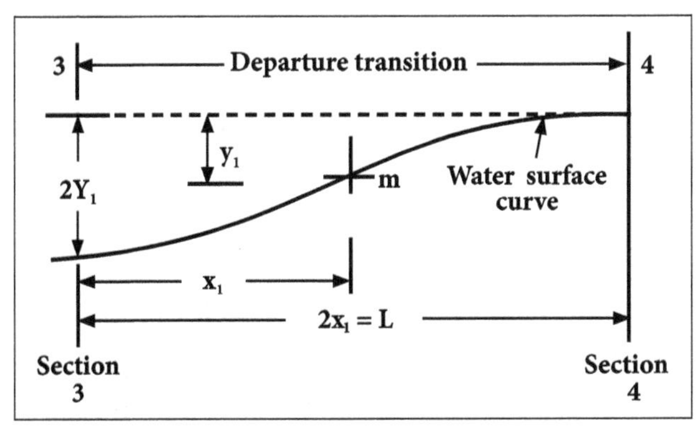

It may be seen from the above figures.

$$L = \text{Length of transition} = 2x_1.$$

$2y_1$= Total drop or rise in the water surface. The point m is the midpoint of transition length and is situated to divide the total drop as well as length equally.

Taking water surface at the section point as the origin equation of the parabola is given by,

$$C = cx_2$$

Substituting the known values of y_1 and x_1

$$C = y_1 / x_2$$

With this value of c parabolic water surface curves can be plotted starting from section points which represent the origin.

The equation is to be used for plotting now reduces to,

$$y = (y_1 / x_{12}).x_2$$

Thus, water surface profile can be plotted.

Step 6: Velocity and area of flow at various points can be determined.

(i) Velocity head at any point is given by the difference between the T.E.L. and the water surface.

Velocity head hv = TEL – WS line

Also,

$$v_2hv = v_2 / 2g$$

So velocity (V) at each point = $\sqrt{2g.hv}$

(ii) The area of flow at any point may now be obtained by using the simple formula

$$A = Q/V$$

With known values of A and D other dimensions of the trapezoidal channel can be calculated using the following formula,

$$A = BD = SD_2$$

Where B is bed width and S : 1, i.e. (H: V) is the side slope.

In case of flared wing walls, the side slopes are gradually brought to vertical from an initial slope. The value of side slope at any intermediate section in the transition length may be interpolated in proportion to the length of the transition achieved up to that point.

Mitra's Hyperbolic Transition Method

This method is based on the principle that:

- Along with the discharge, the depth of flow in the canal is also a constant.
- Rate of change of velocity per unit length of the transition is also constant throughout the length of transition.

From the above design procedure outlined figure it can be seen that:

B_o = Normal bed width of canal.

B_t = Bed width in the throat or trough.

B_x = Width at any distance x from the extremity of the trough.

L = Total length of the transition.

Also let V_o, V_t and V_x be the velocities at the corresponding sections.

From second principle,

$$\frac{V_t - V_x}{x} = \frac{V_t - V_o}{L} \quad \ldots(1)$$

From first principle,

$$Q = V_t \times B_t \times D$$
$$= V_x \times B_x \times D = V_o \times B_o \times D$$

Since Q and D are constant,

Let,

$$\frac{Q}{D} = K$$

$$K = V_t \times B_t = V_x \times B_x = V_o \times B_o$$

or,

$$V_t = \frac{K}{B_t}$$

Also let,

$$V_t = \frac{K}{B_x} \text{ and } V_o = \frac{K}{B_o}$$

Substituting these values in equation (1) above,

$$\frac{\left(\frac{K}{B_t}-\frac{K}{B_x}\right)}{x}=\frac{\left(\frac{K}{B_t}-\frac{K}{B_o}\right)}{L}$$

or,

$$B_x = \frac{B_0 B_t L}{L\,B_o-(B_o-B_t)x}$$

Thus B_x gives bed width at any distance x from the throat.

Chaturvedi's Semi-Cubical Parabolic Transition Method

It states that,

$$x = \frac{LB_0^{3/2}}{B_0^{3/2}-B_t^{3/2}}\left\{1-\left(\frac{B_t}{B_x}\right)^{3/2}\right\}$$

Design Principles for Siphon Aqueduct

It is clear that the siphon aqueducts are basically different from ordinary aqueducts. As such criteria for aqueduct design is not sufficient in the design of siphon aqueducts.

In addition to the above considerations the following criteria should be adopted while designing the siphon aqueducts:

1. Discharge through Siphon Barrel

The head which causes the flow (it also represents head loss in barrel) through the inverted siphon barrel may be obtained from Unwin's formula

$$h=\left(1+f_1+f_2\frac{L}{R}\right)\frac{V^2}{2g}-\frac{V_a^2}{2g}$$

Where,

h - The head causing flow, it is the loss of head in the barrel in m.

L - The length of barrel in m.

R - Hydraulic mean radius of the barrel in m.

V - Velocity of flow through the barrel in m/sec.

V_a - Velocity of the approach in m/sec, it is generally neglected.

f_1 - Coefficient for loss of head at the entry and generally taken as 0.505.

f_2 - Coefficient which accounts for the friction in the barrel.

$$f_2 = a\left(1 + .305\frac{b}{R}\right)$$

Where a and b are constants.

Following Table gives values of a and b for different surfaces:

Type of surface	a	b
Smooth iron pipes	0.00497	0.084
Incrusted iron pipes	0.00996	0.054
Smooth cement plaster	0.00316	0.10
Ashlar brick work	0.00401	0.23
Rubble masonry or stone pitching	0.0050	7 0.83

When the velocity of approach is small, the term representing the loss of head in friction in the barrel $\left(\text{i.e.,} f_2 . \frac{L}{R} . \frac{V^2}{2g}\right)$ may be neglected.

Velocity of flow through the barrel is generally limited to 2 to 3 m/sec.

Thus, All the values are known head loss in barrel or head causing flow may be calculated. This value when added to the High Flood Level (HFL) on d/s of the aqueduct gives u/s HFL.

Adding these free board to u/s HFL we can get the top of river protection works like guide bunds and marginal bunds.

2. Uplift Pressure on the Roof of the Barrel

As the barrel runs full during floods, there exists positive pressure in the barrel. Due to the positive pressure in the barrel, the roof is subjected to uplift pressure. The uplift pressure diagram for the roof can be drawn knowing the pressure head on the u/s and d/s side of the barrel.

Pressure head on the d/s side of the barrel is equal to the height of the water level above the bottom of the roof. Pressure head on the u/s side may be obtained by adding the loss of head in barrel to the pressure head on d/s side. The loss of head may be obtained from the Unwin's formula. Figure below shows the profile of hydraulic gradient line that now exist. It can be seen that maximum uplift pressure occurs at the u/s end of the barrel roof.

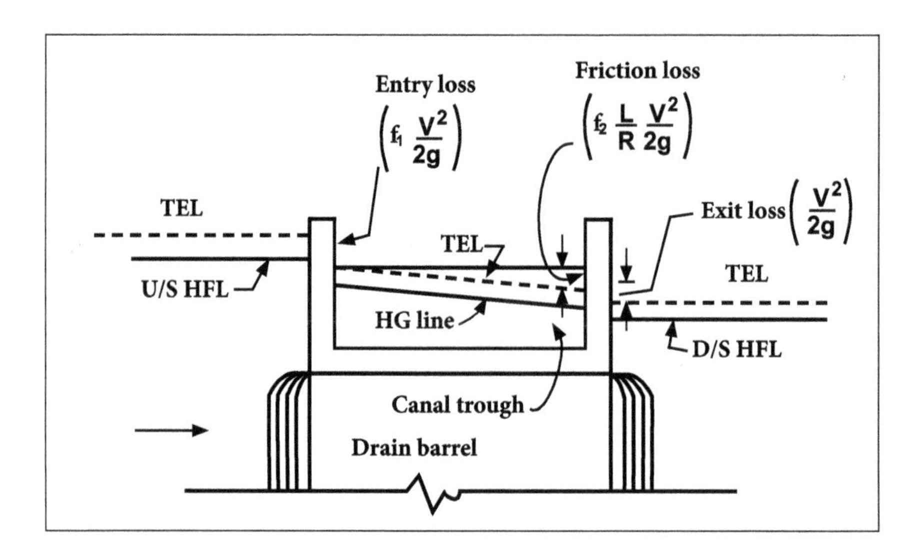

While designing the trough it is necessary to consider two extreme conditions:

- The barrel runs full during maximum flood and there is no water in the canal trough. This condition gives maximum uplift pressure acting on the trough.
- The canal trough is carrying the full discharge but the barrel does not run full and thus there is no uplift on the roof of the barrel.

In order to limit the thickness of the trough it is advisable to provide reinforced concrete roof with reinforcement at the bottom to take the load of the canal trough and the reinforcement at top to resist uplift pressure by bending.

3. Uplift Pressure on the Floor of the Barrel

Unlike other hydraulic structures aqueducts are subjected to two different types of uplift pressures from the two different sources. They are the following:

(i) Static Uplift Pressure Due to Rise in Water Table

The water table rises many times up to bed level of drain. Specifically in case of siphon aqueduct whose floor bed is depressed below bed of the drainage, the static uplift pressure acts on floor bed. The uplift pressure is similar to the difference of the bed level of the drain and that of floor level of the barrel.

(ii) Uplift Pressure Due to Seepage of Canal Water to the Drain

Since there exists the difference of level between canal water level and the drainage water level seepage flow takes place where conditions are desirable. The seepage head is maximum when the canal runs with full capacity and there is no flow in the drain below.

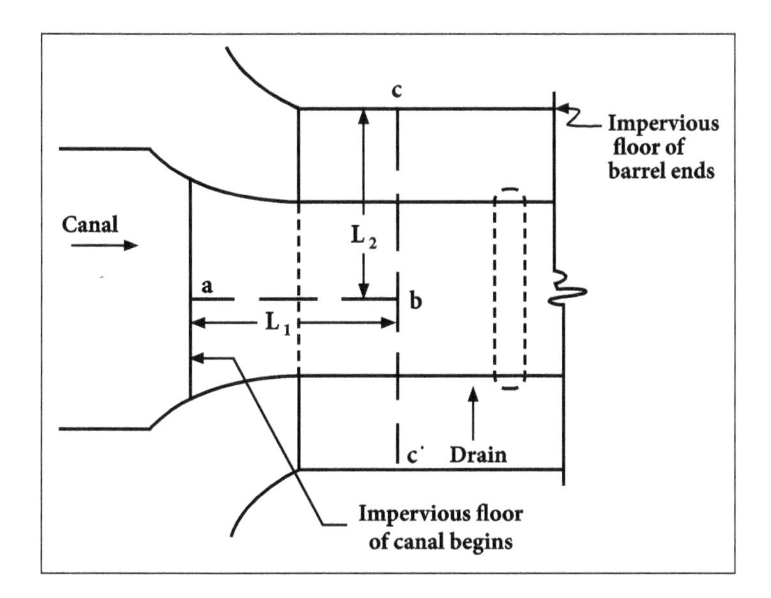

As shown in given Figure below, seepage flow is not simple but flow pattern is found three dimensional everywhere. The seepage flow starts from either sides of impervious canal trough bed and reappears on either sides of the impervious barrel floor in the drain.

Since no approximation to two dimensional flow is possible, Khosla's theory cannot be applied strictly. Solution by intricate "method of relaxations" is still possible but it is too elaborate. For design purposes Bligh's creep theory principle explained below can be applied. For major works, however, it is essential to check the results of the preliminary design so obtained by the model studies.

Taking the case of the first barrel where seepage will be maximum, total creep length = (creep length a-b) + (creep length b-c),

$$L = L_1 + L_2$$

Total seepage head = canal FSL – d/s bed level of drain = Hs

Residual seepage head at b = -Hs/L x L_2

Total residual seepage head at b can be considered to design the thickness of the entire floor of all the barrels. The floor thickness of the barrel is in fact designed considering total uplift pressure created by static uplift condition and canal seepage flow mentioned above.

In order to limit the thickness of the floor RCC construction can be adopted as the part of the pressure is resisted by weight of floor and remaining by bending strength of the floor. In such arrangement the pressure gets transferred to the piers and is resisted by the entire weight of the superstructure.

When it is seen that uplift pressure is very high it can be reduced by providing suitable safe guards. They are:

- Increasing the length of the impervious floor of the canal bed so that creep length is also increased.
- Providing drainage holes or relief holes in the floor of the barrel in conjunction with inverted filter below the floor. To avoid choking of the relief holes and filter below the drain silt relief holes must be provided with flap valves.

Super Passage

The hydraulic structure in which the drainage is passing over the irrigation canal is termed as super passage. This structure is suitable when the bed level of drainage is above the flood surface level of the canal. The water of the canal passes clearly below the drainage.

- A super passage is same to an aqueduct, except in this case the drain is over the canal.
- The FSL of the canal is lower than the underside of the trough carrying drainage water. Thus, the canal water runs under the gravity.
- Reverse of an aqueduct.

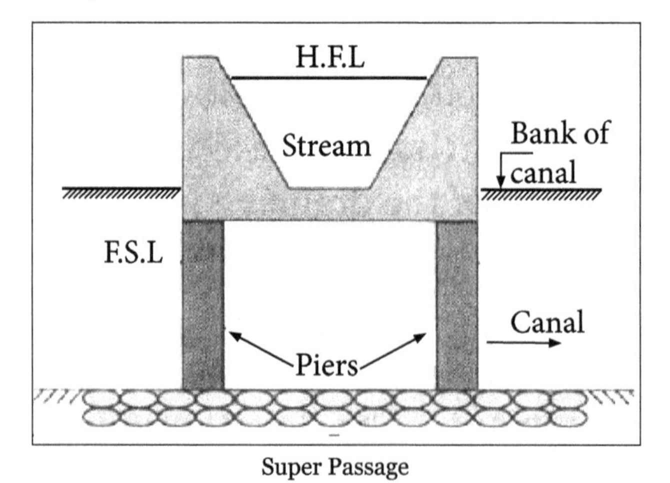

Super Passage

2.5 Diversion Head Works

It is constructed at the head of canal and diverts the river water towards the canal to ensure the regulated and continuous supply of the silt-free water with certain minimum head into the canal.

Objective of these Diversion Head Works

- To rise the water level at the head of the canal.
- To form the storage by constructing dykes on both the river banks such that water is available throughout the whole year.
- To control the entry of silt into canal and also to control the deposition of the silt at the head of the canal.
- To control the fluctuation of the water level in the river during different seasons of the year.

Selection of the Site for Diversion Head Works

At the site, the river must be straight and narrow, the river banks must be well defined. The valuable land must not be submerged when the weir or barrage is constructed. Elevation of the site must be much higher than the area that is to be irrigated.

The site must be easily accessible either by roads or railways. The materials of construction must be available in vicinity of the site. The site must not be far away from the command area of the project, in order to avoid transmission loss.

2.5.1 Weirs and Barrages

It is the barrier built across river to raise the water level on the upstream side of obstruction to feed the main canal.

The ponding of the water may be achieved either by a raised crest across the river or by the raised crest supplemented by gates or the shutters, working over crest.

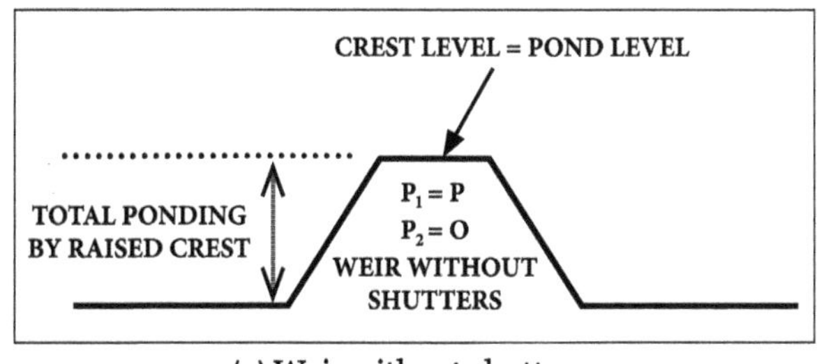

(a) Weir without shutters.

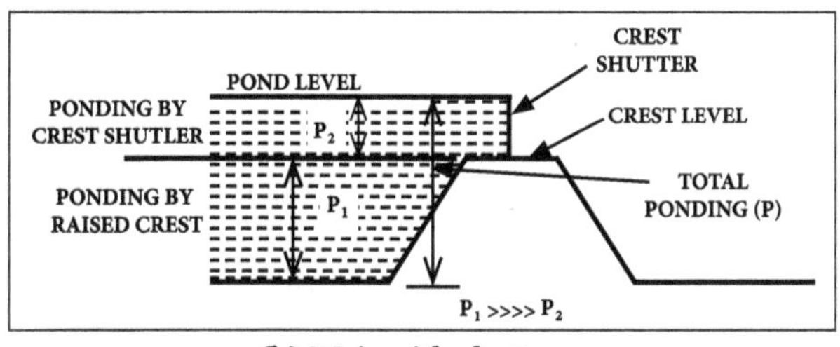

(b) Weir with shutters.

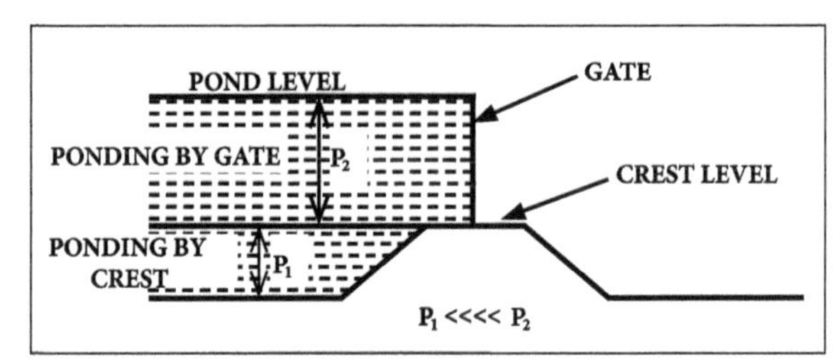

(c) Barrage with a small raised crest.

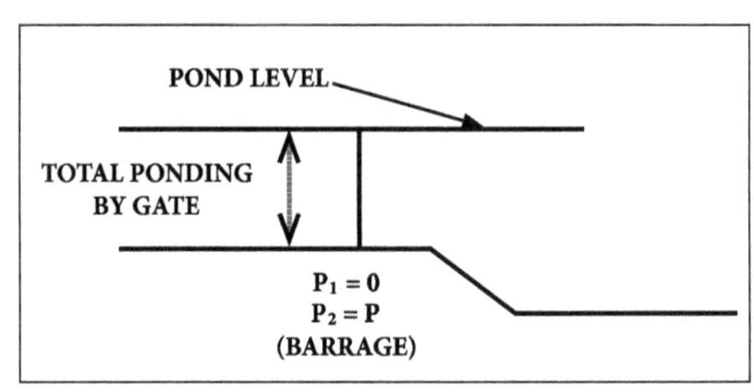

(d) Barrage without any raised crest.

Weir

If major part or entire ponding of the water is achieved by a raised crest and a smaller part or nil part of it, is achieved by means of shutters, then this barrier is called as a weir.

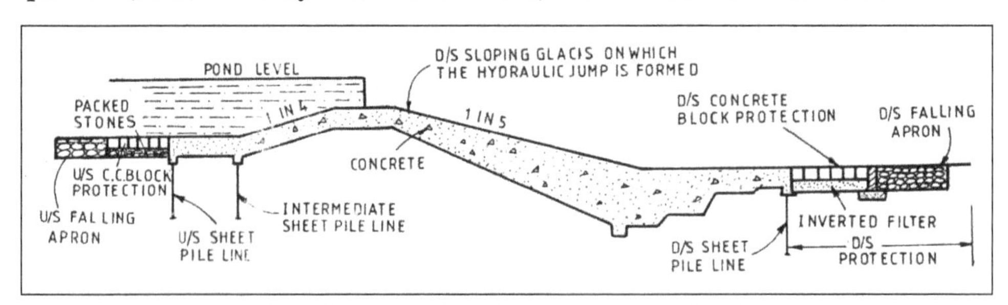

A typical cross-section of a modern concrete weir.

Gravity and Non-Gravity Weirs

When the weight of weir balances the uplift pressure caused by head of the water seeping below weir, it is known as a gravity weir. On the other hand, if the weir floor is designed continuous with the divide piers as reinforced structure, in a manner that the weight of concrete slab together with the weight of divide piers keeping the structure safe against the uplift, then that structure can be called as a non-gravity weir.

In the latter case, RCC is to be used in place of the brick piers.

Considerable savings can be obtained, as weight of the floor may be much lesser than what is needed in gravity weir.

Barrage

If most of the ponding is done by gates and the smaller or nil part of it is done by the raised crest, then the barrier is referred as a barrage or a river regulator.

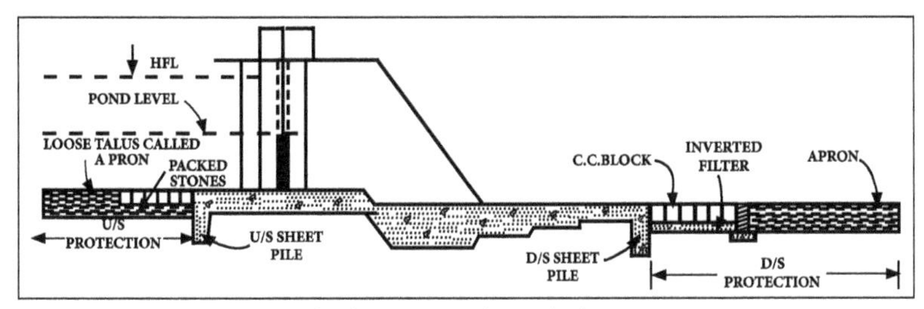

A typical cross-section of a barrage.

The only difference between a weir and a barrage is of the gates that are the flow in which the barrage is regulated by gates and that in weirs, by its crest height. Barrages are costlier than weirs.

Weirs and barrages are constructed mostly in the plain areas. The heading up of water is affected by gates that are put across the river. The crest level in the barrage is kept at the low level. During flood, gates are raised to clear the high flood level. As a result there is less silting and provide better regulation and control than the weir.

Components of Barrage

Main Barrage Portion

Main body of the barrage has normal RCC slab in which it supports the steel gate. In the X-Section it consists of: Upstream concrete floor, to lengthen the path of the seepage and to project the middle portion where the pier, gates and bridge are located.

It has crest at the required height above the floor on which the gates rest in their closed position and the Upstream glacis of suitable slope and shape. This joins the crest to the downstream floor level. The hydraulic jump forms on the glacis since it is more stable than on the horizontal floor, this eventually reduces the length of the concrete work on the downstream side.

Downstream floor is built with the help of concrete and is constructed so as to contain the hydraulic jump. Therefore it takes care of turbulence which would otherwise cause erosion. It is also provided with the friction blocks of suitable shape and at a distance determined through the hydraulic model experiment to increase the friction and to destroy the residual kinetic energy.

Divide Wall

A wall constructed at right angle to the axis of the weir separate the weir proper from the under sluices. It extends upstream beyond the beginning of canal HR. Downstream it extends up to the end of loose protection of under sluices. This is to cover the hydraulic jump and resulting turbulence.

The Fish Ladder

For movement of fish:

- Difference of level on upstream and downstream sides on weir is split up into water steps by means of the baffle walls that is constructed across the inclined chute of fish ladder.
- Velocity in the chute should not be more than 3m/s.
- Grooved gate at the upstream and the downstream - for effective control.
- Optimum velocity 6-8 ft/s.

Sheet Piles

It is made of mild steel, each portion being 1/2' to 2' in width and 1/2" thick and of the required length, having groove to link with the other sheet piles.

Upstream Piles

It is situated at upstream end of upstream concrete floor driven into soil beyond the maximum possible scour that shall occur.

Functions

- Protects barrage structure from scour.
- Reduces the uplift pressure on barrage.
- Holds the sand compacted and densified between two sheet piles and thereby increases the bearing capacity when barrage floor is designed as raft.

Intermediate Sheet Piles

It is situated at the end of upstream and downstream glacis. Protection to main structure of barrage in the event of upstream and downstream sheet piles collapsing due to the advancing scour or undermining. They also help lengthen the seepage path and reduce the uplift pressure.

Downstream Sheet Piles

It is placed at the end of the downstream concrete floor. Their main function is to check exit gradient. Their depth must be greater than the possible scour.

Inverted Filter

It is provided between downstream sheet piles and flexible protection. Typically 6" sand, 9" coarse sand and 9" gravel. Filters shall vary with size of the particles forming the river bed. It is protected by placing over the concrete blocks of sufficient weight and size. Slits are left between the blocks to allow the water to escape.

The length must be 2x downstream depth of sheet.

Functions

- Checks the escape of fine soil particles in the seepage water.
- Flexible apron.
- Placed on the downstream of the filter.
- Consists of boulder large enough not to be washed away by the highest likely velocity.

The protection provided is enough as to cover the slope of scour of 1 1/2 x depth of scour as the upstream side of 2x depth of the scour on the downstream side at the slope of 3.

The Under Sluices

- Scouring sluices.
- Maintaining a deep channel in front of the Head regulator on the downstream side.

Functions

As the bed of under sluice is not lower than rest of the weir, most of the day, flow unit will flow toward this pocket → easy diversion to the channel through the Head regulator.

- Control the silt entry into channel.
- Scour the silt.
- High velocity currents due to high differential head.
- Passes the low floods without dropping.

The shutter of the main weir, the raising of which entails good deal of labor and time.

Capacity of the Under Sluices

For sufficient scouring capacity, its discharging capacity must be at least double the canal discharge. Must be able to pass the dry weather flow and the low flood, without dropping the weir shutter. It is capable of discharging 10 to 15% of high flood discharge.

Choice between a Weir and a Barrage

The choice between a weir and a barrage is largely governed by the cost and convenience in working.

A shuttered weir could be relatively cheaper but will lack the effective control possible in the case of a barrage.

A barrage type construction may be easily supplemented with a roadway across the river at a small additional cost. Barrages are almost invariably constructed now-a-days on all the important rivers.

2.5.2 Types of Weirs and Barrages

Types of weirs:

- Masonry weirs with vertical drop.
- Rock-fill weirs with sloping aprons.
- Concrete weirs with sloping glacis.

1. Masonry Weirs with Vertical Drop

Masonry weir wall is constructed over impervious floor. Cut-off walls are present at both ends of the floor. Sheet piles are provided below the cut off walls. The crest shutters are provided to raise the water level, if needed. The shutters are dropped down during the flood.

The masonry weir wall can be vertical on both face and sloping on both face or vertical on the downstream face and sloping in the upstream face.

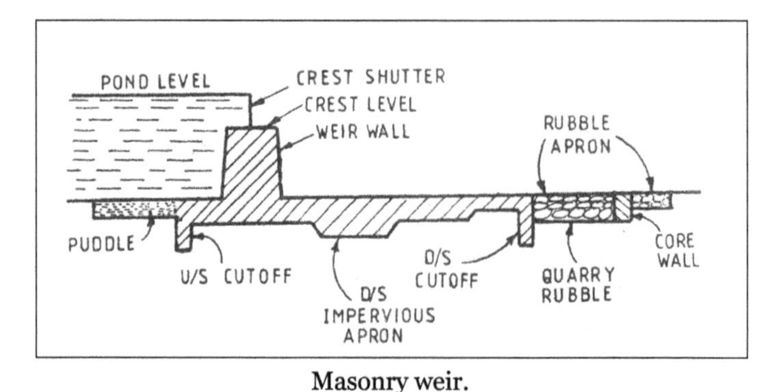

Masonry weir.

2. Rock-Fill Weirs with Sloping Aprons

It consists of a masonry breast wall which is provided with the adjustable crest shutter. The upstream rock-fill portion is constructed with the boulders forming a slope of 1 in 4. The boulders are grouted with the cement mortar.

The downstream sloping apron consists of the core walls. The intermediate spaces between the core walls are filled up with the boulders maintaining a slope of 1 in 20. The boulders are grouted properly with the cement mortar.

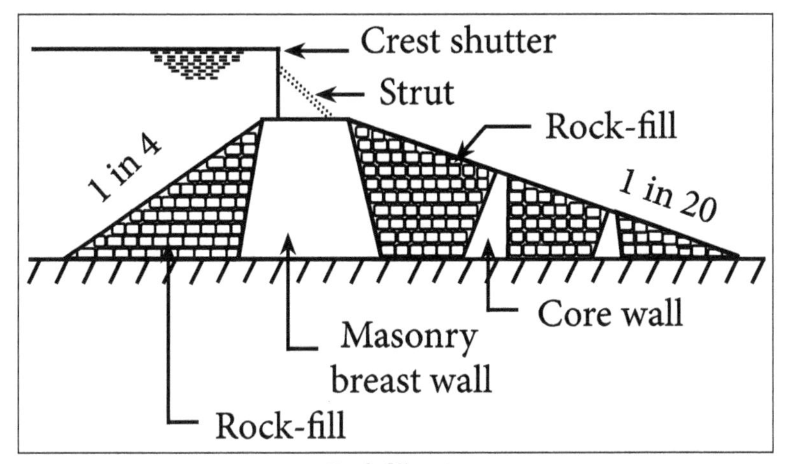

Rock fill weir.

3. Concrete Weir

Now-a-days, the weirs are constructed with the help of reinforced cement concrete. The impervious floor and the weir are both made monolithic. The cut off walls are provided at the upstream and the downstream end of the floor and at the toe of the weir. Sheet piles are provided below the cut-off walls. The crest shutters are provided which are dropped down during the flood.

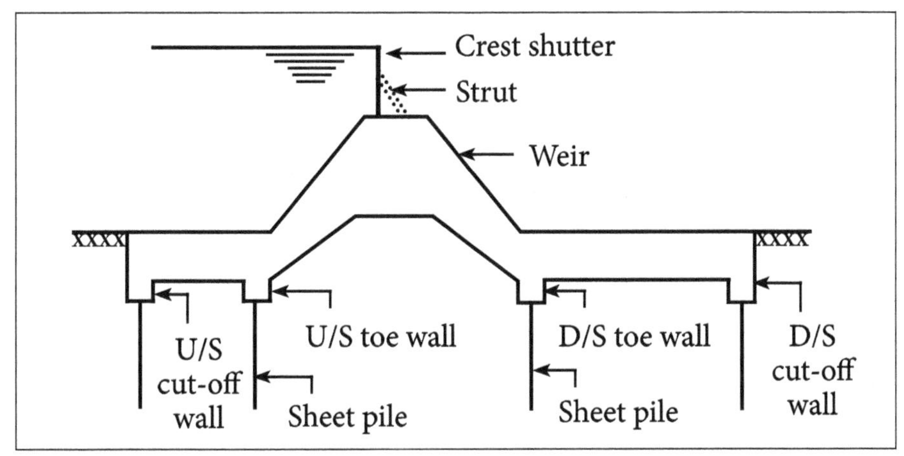

Concrete weir.

2.5.3 Layout of a Diversion Head Works

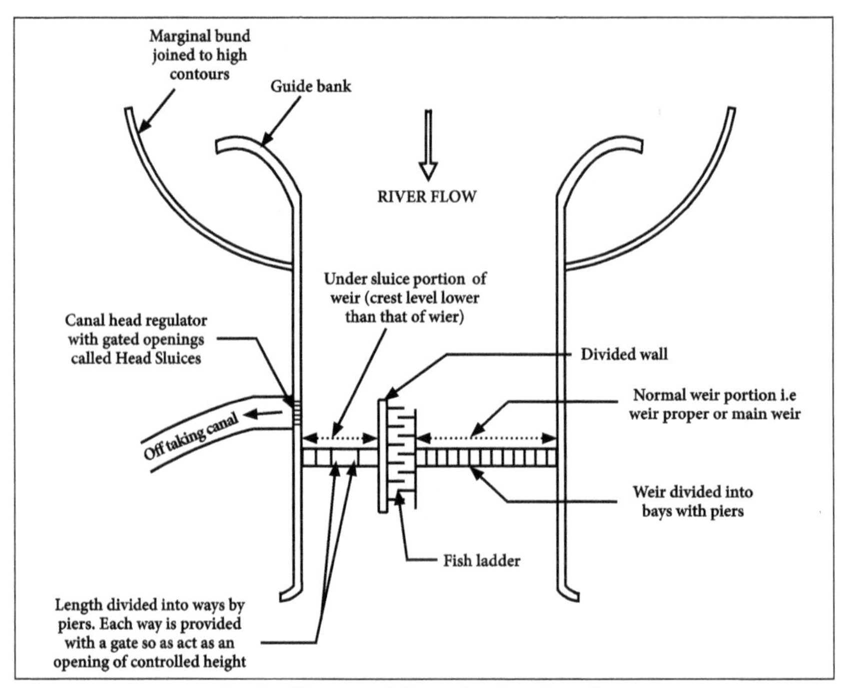

Typical layout of diversion head works.

A typical layout of a canal head-works is shown in figure. This head-works consists of:

- Weir proper.
- Under-sluices.
- Divide wall River.
- Training works.
- Fish Ladder.
- Canal Head Regulator.

2.6 Different Components of a Diversion Head Works

Weir Proper

It is a barrier built across the river. It's aim is to raise the water level to feed the canal.

Under-Sluices

The under sluices are the openings given at the base of weir or barrage. These openings are provided with the adjustable gates. Usually, the gates are kept closed. The crest of under sluice portion of the weir is kept at a lower level than the crest of the normal portion of the weir. The suspended silt goes on depositing in front of canal head regulator.

When silt deposition becomes appreciable the gates are opened and the deposited silt is loosened with an agitator mounting on a boat. The muddy water flows towards the downstream through the scouring sluices. The gates are then closed. Whereas, at the period of flood, the gates are kept opened. The main functions of under-sluices are:

- To maintain a well-defined deep channel approaching the canal head regulator.
- To ensure a easy diversion of water into the canal through the canal head regulator even during low flow.
- To control the entry of silt into the canal.
- To help the scouring and the silt deposited over the under-sluice floor. To also help removing it towards the downstream side.
- To help passing the low floods without dropping the shutters of weir.

The Divide Wall

The divide wall is the masonry or concrete wall constructed at the right angle to the axis of the weir. The divide wall is extending on the upstream side beyond the beginning of canal head regulator and also on the downstream side. It extends up to the end of the loose protection of the under-sluices. The divide wall is a really long wall built at the right angles in weir or barrage, it can be constructed with the stone masonry or the cement concrete.

On the upstream side, the wall is extended just to cover the canal head regulator and on the downstream side, it is extended up to launching apron.

The main functions of divide walls:

- It separates the 'under-sluices' with the lower crest level from the 'weir proper' with the higher crest level.
- It helps in providing a comparatively less turbulent pocket near canal head regulator, resulting in the deposition of silt in this pocket and hence, to help in the silt-free water to enter into the canal.
- It helps to keep cross-current away from the weir.

Fish Ladder

- It is a device by which energy flow can be dissipated in a manner so as to provide smooth flow at essentially low velocity, not exceeding from 3 to 3.5 m/s.
- A narrow opening that included suitable baffles or staggering devices in it is given adjacent to the divide wall.
- The fish ladder is provided just by the side of divide wall for the free movement of the fish. Rivers are important source of fish.
- There are several types of fish in the river. The nature of each fish varies from type to type. But in general, the tendency of fish is to move from the upstream to the downstream in winters and from the downstream to the upstream in monsoons.
- This movement is very essential for their survival. Due to the construction of weir or barrage, this movement gets obstructed and they are detrimental to the fishes.
- In the fish ladder, fable walls are constructed in a zigzag manner so that the velocity of flow within ladder does not exceed 3 m/sec.
- The width, length and the height of the fish ladder depends on the nature of the river and the type of the weir or barrage.

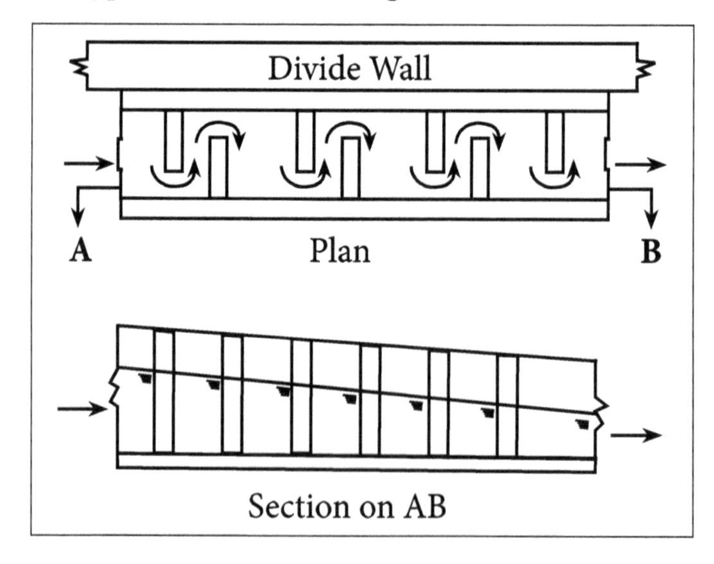

Canal Head Regulator or Head Sluices

The structure which is constructed at the head of the canal in order to regulate flow of water is referred as canal head regulator. It has a number of piers which divide the total width of the canal into number of spans which are termed as bays. The piers consist of number tiers on which the adjustable gates are present.

The gates are operated from the top by a suitable mechanical device. A platform is provided on the top of the piers for facility of gate operation. Again some piers are constructed on downstream side of canal head to support the roadway.

Head Regulator

- It regulates the water supply that enters into the canal.
- It controls the entry of silt in the canal.
- It prevents river-floods from entering into canal.

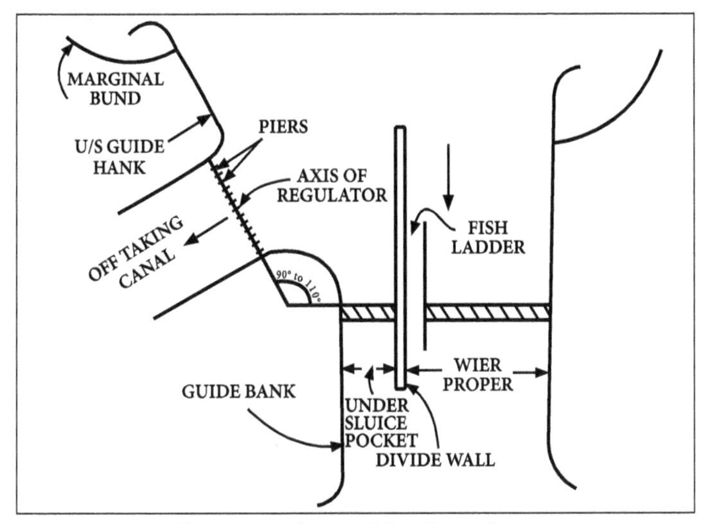

Alignment of a canal head regulator.

The water from the under-sluice pocket is made to enter regulator bays, such that to pass the full supply discharge into canal. The maximum height of such gated openings, known as head sluices will be equal to difference of the Pond Level and the Crest Level of the regulator.

The entry of silt into canal is controlled by keeping crest of head regulator by about 1.2 to 1.5 meters higher than the crest of under-sluices. When a silt-excluder is provided, the regulator crest is moreover raised by about 0.6 to 0.7 meter. Silt gets deposited in the pocket and only clear water enters into regulator bays. The deposited silt may be easily scoured out regularly and removed through under-sluice openings.

2.7 Design of Weirs and Barrages

Bligh's Creep Theory

According to the Bligh's Theory, the percolating water follows the outline of the base of foundation of the hydraulic structure. In other words, water creeps along the bottom contour of structure. The length of the path hence traversed by water is known the length of the creep. Moreover, it is assumed in this theory, that the loss of head is proportional to the length of the creep.

If HL is total head loss between the upstream and the downstream and L is length of creep, then the loss of head per unit of creep length is known as the hydraulic gradient. Further, Bligh makes no distinction between horizontal and vertical creep.

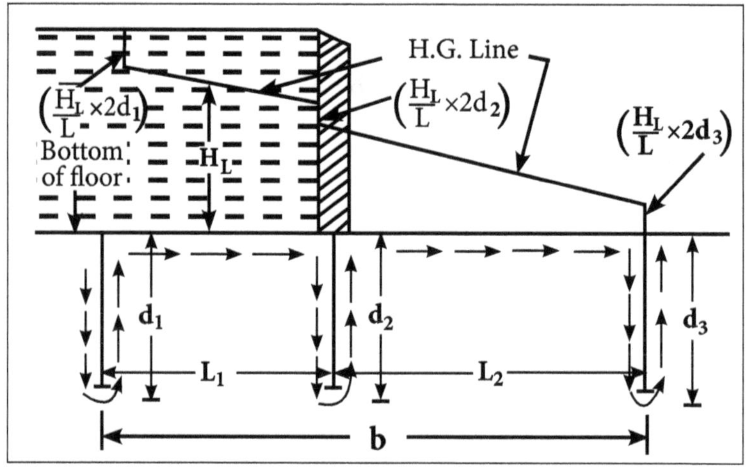

Bligh's creep.

Bligh called the loss of head per unit length of creep as the Percolation coefficient. The reciprocal, (L/H) of the percolation coefficient is called as the coefficient of creep C.

Assumptions

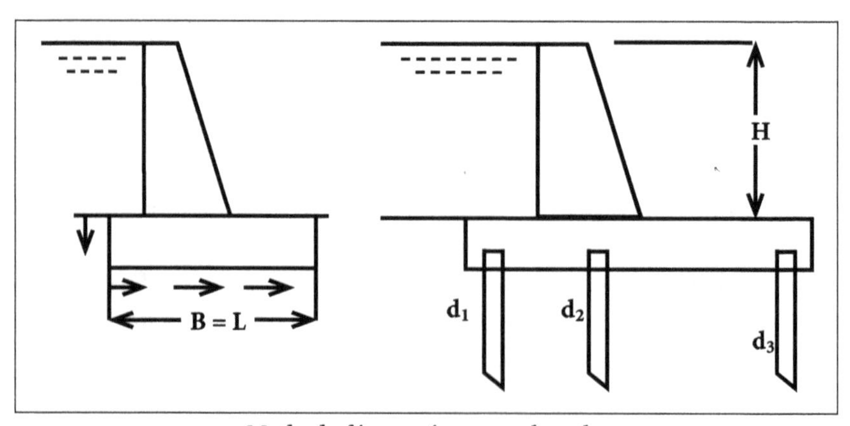

Method of increasing creep length.

Hydraulic slope or gradient is constant throughout the impervious length of the apron. The percolating water creep along the contact of the base profile of the apron with the sub soil losing head en route, proportional to the length of its travel. The length is called the creep length. It is the sum of horizontal and vertical creep. Stoppage of the percolation by the cut off is possible only if it extends up to the impermeable soil strata.

Consider a section a shown in Figure above. Let H_L be the difference of water levels between the upstream and the downstream ends. Water will seep along the bottom contour as shown by the arrows. It starts percolating at A and emerges at B. The total length of the creep is given by,

$$L = d_1 + d_1 + L_1 + d_2 + d_2 + L_2 + d_3 + d_3$$

$$= (L_1 + L_2) + 2(d_1 + d_2 + d_3)$$

$$= b + 2(d_1 + d_2 + d_3)$$

Head loss per unit length or gradient,

$$C' \frac{H}{L} = \frac{H}{\left[B + 2\left(d_1 + d_2 + d_3\right)\right]}$$

Head losses equal to $\left(\frac{H_L}{L} \times 2d_1\right), \left(\frac{H_L}{L} \times 2d_2\right), \left(\frac{H_L}{L} \times 2d_3\right)$ will occur respectively, in the planes of the three vertical cut offs. The hydraulic gradient line may then be drawn as shown in figure above.

1. Safety Against Piping or Undermining

According to Bligh, the safety against piping may be ensured by providing the sufficient creep length, given by $L = C.H_L$, where C is the Bligh's Coefficient for the soil. Different values of C for different types of soils are tabulated in Table below:

S. No	Type of soil	Value of C	Safe hydraulic gradient should be less than
1	Fine micaceous sand	15	1/15
2	Coarse grained sand	12	1/12
3	Sand mixed with the boulder and gravel, and for loam	5 to 9	1/5 to 1/9
4	Light sand and mud	8	1/8

Note: The hydraulic gradient i.e. H_L/L is then equal to 1/C. Hence, it may be stated that the hydraulic gradient should be kept under a safe limit in order to ensure safety against piping.

2. Safety Against Uplift Pressure

The ordinates of H.G line above the bottom of the floor represent residual uplift water head at each point. Say for example, if at any point, ordinate of H.G line above the bottom of the floor is 1 m, then 1 m head of water may act as uplift at that point. If h′ meters is this ordinate, then the water pressure equal to h′ meters will act at this point and it has to be counterbalanced by the weight of the floor of thickness say t.

Uplift pressure $= \gamma w \times h'$ [where γw is the unit weight of water]

Downward pressure $= (\gamma w \times G).t$ [Where G is specific gravity of the floor material]

For equilibrium,

$$\gamma w \times h' = \gamma w \times G.\ T$$

$$h' = G \times t$$

For equilibrium,

$$\Rightarrow h' = G \times t$$

Subtracting t on both sides, we get,

$$(h' - t) = (G\,x\,t - t) = t(G - 1)$$

$$t = \left(\frac{h' - t}{G - 1}\right) = \left(\frac{h}{G - 1}\right)$$

Where, $h' - t = h =$ Ordinate of the H.G line above the top of the floor G – 1 = Submerged specific gravity of the floor material.

Design of Impervious Floor for Sub Surface Flow

It is directly dependent on the possibilities of percolation in porous soil on which the floor is built. Water from the upstream percolates and creeps slowly through the weir base and the subsoil below it.

The head lost by the creeping water is proportional to the distance it travels along the base of the weir profile. The creep length should be made as big as possible so as to prevent the piping action. This shall be achieved by providing the deep vertical cut-offs or sheet piles.

Limitations of the Bligh's Theory

This theory made no distinction between the horizontal and the vertical creep:

- It did not explain the idea of exit gradient - safety against the undermining cannot be simply obtained by considering a flat average gradient but by keeping this gradient will be critical.
- No distinction between the outer and the inner faces of the sheet piles or the intermediate sheet piles, whereas from investigation it is clear, that outer faces of the end sheet piles are much effective than the inner ones.

- Losses of head does not take place in the same proportions as creep length. The uplift pressure distribution is not linear but it follows a sine curve.
- In case of two piles the width between must be greater than twice the head or the piles.

Lane's Weighted Creep Theory

Bligh, in his theory, had calculated the length of the creep, by simply adding the horizontal creep length and the vertical creep length, thereby making no distinction between the two creeps.

However, Lane, on the basis of his analysis carried out on about 200 dams all over the world, stipulated that the horizontal creep is less effective in reducing uplift (or in causing loss of head) than the vertical creep. Therefore, he suggested a weightage factor of 1/3 for the horizontal creep, as against 1.0 for the vertical creep.

Thus in above figure, the total Lane's creep length (L_1) is given by,

$$\begin{aligned} L_1 &= (d_1 + d_1) + (1/3)\ L_1 + (d_2 + d_2) + (1/3)\ L_2 + (d_3 + d_3) \\ &= (1/3)\ (L_1 + L_2) + 2(d_1 + d_2 + d_3) \\ &= (1/3)b + 2(d_1 + d_2 + d_3) \end{aligned}$$

To ensure safety against piping, according to this theory, the creep length L_1 must not be less than C_1H_L, where H_L is the head causing flow and C_1 is Lane's creep coefficient given in table.

Values of Lane's Safe Hydraulic Gradient for different types of Soils:

SL. No.	Type of Soil	Value of Lane's Coefficient C_1	Safe Lane's Hydraulic gradient should be less than
1	Very fine sand or silt	8.5	1/8.5
2	Fine sand	7.0	1/7
3	Coarse sand	5.0	1/5
4	Gravel and sand	3.5 to 3.0	1/3.5 to 1/3
5	Boulders, gravels and sand	2.5 to 3.0	1/2.5 to 1/3
6	Clayey soils	3.0 to 1.6	1/3 to 1/1.6

2.7.1 Khosla's Theory, Khosla's Method of Independent Variables, Exit Gradient

The main principles of this theory are summarized below:

(a) The seepage water does not creep along bottom contour of pucca flood as started by the Bligh, but on the other hand, this water moves along the set of stream-lines.

This steady seepage in a vertical plane for a homogeneous soil may be expressed by Laplacian equation:

$$\frac{d^2\phi}{dx^2}+\frac{d^2\phi}{dz^2}$$

Where,

φ = Flow potential = Kh.

K = The co-efficient of permeability of soil as defined by Darcy's law.

h = The residual head at any point within the soil.

The above equation represents sets of two individual curves intersecting each other orthogonally. The resultant flow diagram showing both of the curves is called a Flow Net.

Stream Lines: The streamlines represent the paths along which the water flows through the sub-soil. Every particle entering the soil at a given point upstream of the work, will trace out its own path and may represent a streamline. The first streamline follows the bottom contour of the works and is same as Bligh's path of creep. The remaining streamlines follows smooth curves transiting slowly from outline of the foundation to a semi-ellipse, as shown.

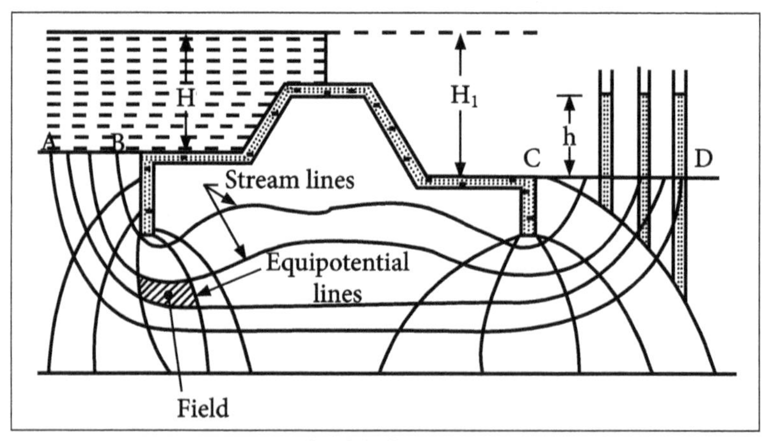

Khosla's flow net.

Equipotential Lines: It treats the downstream bed as datum and assuming no water on the downstream side, it may be easily stated that every streamline possesses a head similar to H_1 while entering into the soil and when it emerges at the down-stream end into the atmosphere, its head is zero. Therefore, the head H1 is entirely lost during the passage of water along the streamlines.

Further, at every intermediate point in its path, there is a certain residual head (h) still to be dissipated in remaining length to be traversed to downstream end. This fact is applicable to every streamline and thus, there will be points on different streamlines having same value of residual head h. If such points are joined together, the curve obtained is known as an equipotential line.

Every water particle on line AB having a residual head h = H_1 and on CD having a residual head h= 0 and thus, AB and CD are the equipotential lines. As an equipotential line represents the joining of points of equal residual head, thus if piezometers were installed on an equipotential line, the water will rise in all of them up to same level as shown in figure below.

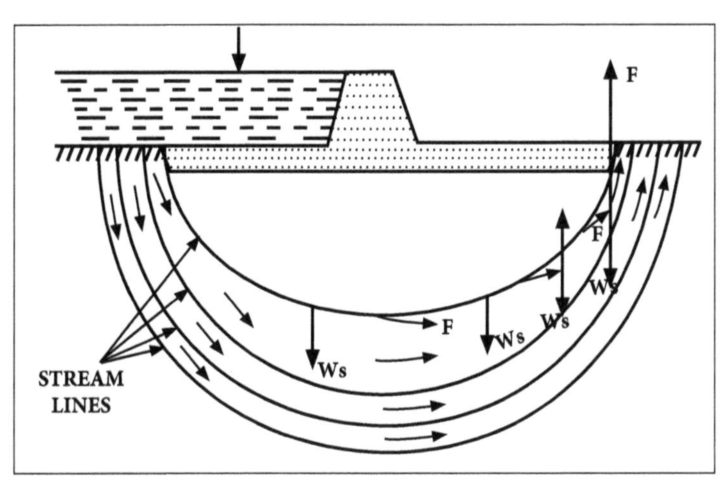

(b) The seepage water exerts a force at each point in the direction of the flow and tangential to the streamlines as shown in figure above. This force (F) has an upward component from the point where streamlines turns upward.

For soil grains to remain stable, the upward component of this force must be counterbalanced by the submerged weight of the soil grain. This force has the maximum disturbing tendency at the exit end, this is because the direction of this force at the exit point is vertically upward and therefore full force acts as its upward component.

For soil grain to remain stable, the submerged weight of the soil grain must be more than this upward disturbing force. The disturbing force at any point is always proportional to the gradient of pressure of water at that point. This gradient of pressure of water at the exit end is known as the exit gradient. In order that the soil particles at exit to remain stable, the upward pressure at exit must be safe. In other words, the exit gradient must be safe.

Critical Exit Gradient

This exit gradient is said to be critical, when upward disturbing force on the grain is just equal to the submerged weight of the grain at the exit. When a factor of safety equal to the 4 to 5 is taken into account, the exit gradient may then be taken as safe.

In other words, an exit gradient that equals to 1/4 to 1/5 of the critical exit gradient is ensured, so as to keep the structure safe against piping. The submerged weight (W_s) of a unit volume of the soil is given as:

$$\gamma_w(1-n)(S_s-1)$$

Where,

γ_w = Unit weight of water.

S_s = Specific gravity of the soil particles.

n = Porosity of the soil material.

For some critical conditions to occur at the exit point.

$$F = W_s$$

Where F is the upward disturbing force on the grain Force.

$$F = \text{pressure gradient at that point} = dp/dl = \gamma_w \times dh/dl$$

After studying a lot of dam failure constructed based on Bligh's theory, Khosla came out with the following outcome:

- From observation of Siphons designed on Bligh's theory, by actual measurement of pressure, with the help of pipes inserted in the floor of two of the siphons. It does not show any relationship with pressure calculated on Bligh's theory.
- The outer faces of end sheet piles were much more effective than the inner ones and the horizontal length of floor.
- The intermediate piles of smaller length were ineffective except for local redistribution of pressure.
- Undermining of floor started from tail end.
- It was essential to have a reasonably deep vertical cut off at downstream end to prevent undermining.
- Khosla and his associates took into account the flow pattern below the impermeable base of hydraulic structure to calculate uplift pressure and exit gradient.
- Starting with a simple case of horizontal flow with negligibly small thickness.
- Seeping water below a hydraulic structure does not follow the bottom profile of the impervious floor as stated by Bligh but each particle traces the path along a series of streamlines.

For the case of two dimensional flows under the straight floor:

- For first flow line AB which touches outline of the floor, pressure shall be determined by putting the different values of x in the equation.
- Slope of Pressure diagram: Having A and B in infinite, thus the floor at A will be theoretically infinite acting downward and that at B may also be infinite acting upward. This will cause sand boiling and so the floor should be depressed or cut off must be provided at downstream end.

Composite Profile

The following specific causes of the general form were considered:

- Straight and horizontal floor of negligible thickness with pile at the either end, upstream or at the downstream end.
- Straight and horizontal floor of negligible thickness with pile at some intermediate point.
- Straight and horizontal floor, depressed below the bed, but with no cut off.

Method of Independent Variable

- Most designs do not confirm to elementary profiles. In actual cases we may have a number of piles at upstream level, downstream level and the intermediate points and the floor may also have some thickness.
- Khosla solves the actual problem by using an empirical method called as the method of independent variables.
- This method consists of breaking up complex profile into a number of simple profiles, each is independently amiable to the mathematical treatment then by applying corrections due to the thickness of slope of floor.
- Hence an example complex profile shown in figure is broken up to the following simple profile and the pressure at Key Points are calculated.
- Straight floor of negligible thickness with pile at the upstream ends.
- Straight floor of negligible thickness with pile at the downstream end.
- Straight floor of negligible thickness with pile at the intermediate points.
- The pressure is obtained at the key points by considering the simple profile.

For the determination of seepage below the foundation of the hydraulic structure developed the method of independent variable is used.

In this method, actual profile of a weir which is complex, is divided into a number simple profiles, each of which can be solved mathematically without difficulty. The most useful profile considered are:

- A straight horizontal floor of negligible thickness provided with a sheet pile at the upstream end or a sheet pile at the downstream end.
- A straight horizontal floor depressed below the bed, but without any vertical cut-off.

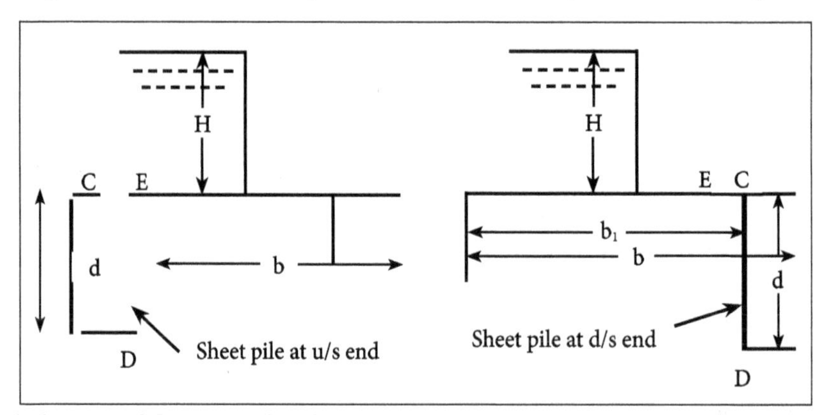

A straight horizontal floor of negligible thickness with a sheet pile at some Intermediate point.

The mathematical solution of the flow-nets of the above profiles have been given in the form of curves. From the curves, percentage pressures at various key points can be determined. The important points to note are:

- Junctions of pile with the floor on either side.
- Bottom point of pile (D).
- Junction of the bottom corners (D, D') in case of the depressed floor.

The percentage pressures at the key points of a simple forms will become valid for any complex profile, provided the following corrections are effected:

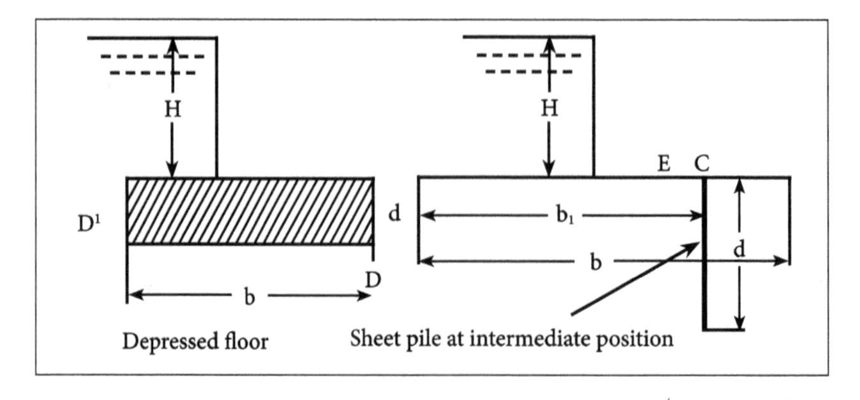

- Corrections for mutual interference of piles.

- Corrections for the thickness of floor.
- Corrections for slope of the floor.

Corrections for Mutual Interference of Piles

Let,

b_1 = Distance between the two piles 1 and 2.

D = The depth of the pile line (2), the influence of which on the neighboring pile (1) of depth d should be determined.

b = Total length of impervious floor.

c = Corrections due to the interference.

The correction is applied as the percentage of the head,

$$C = 19\sqrt{\frac{D}{b'}}\left(\frac{d+D}{b}\right)$$

This correction is positive if the point is considered to be at the rear of interfering pile and negative for the points considered in the forward or flow of direction with the interfering pile.

Corrections for the Floor Thickness

Standard profiles assuming the floors as having negligible thickness. Hence the values of the percentage pressures computed from the curves corresponds to the top levels (E_1^*, C_1^*) of the floor. Moreover, the junction points of the floor and pile are at the bottom of the floor (E_1, C_1).

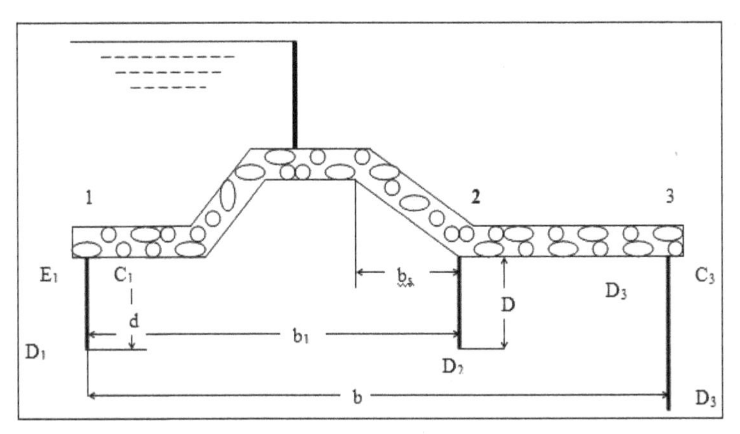

The pressures at an actual points E1 and C1 are interpolated by assuming a straight line variation in the pressures from the points E_1^* to D_1 and from D_1 to C_1.

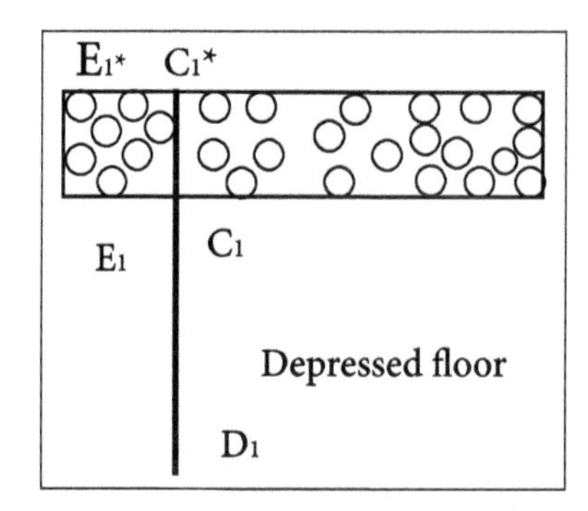

The corrected pressures at E_1 must be less than computed pressure t E_1^*. Hence the correction for the pressure at E_1 will be negative. And so also is for pressure at C_1.

Correction for Slope of Floor

A correction for sloping impervious floor is positive for down slope in the flow direction and negative for the up slope in the direction of flow:

S.No.	Slope = Ver:Horiz	Correction as % of pressure
1	1:1	11.2
2	1:2	6.5
3	1:3	4.5
4	1:4	3.3
5	1:5	2.8
6	1:6	2.5
7	1:7	2.3
8	1:8	2.0

The correction factor should be multiplied by the horizontal length of the slope and divided by the distance between two poles between which the sloping floor exists.

In the diagram above, correction for slope can be applied only to point E_2. As the point E_2 is terminating at descending slope in the direction of flow, the correction will be positive.

The value of correction will be:

$$C.F. \times b_s/b_1$$

Where,

C.F. = The correction factor.

b_s = Horizontal length of the sloping floor.

b_1 = Horizontal distance between pile lines.

Exit Gradient

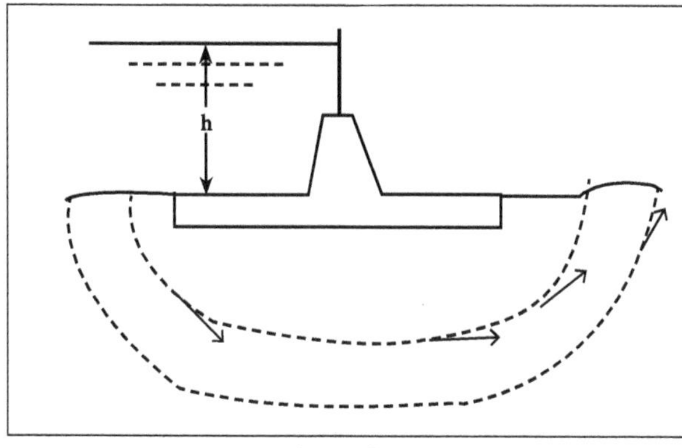

Exit gradient.

Every particle of water when seeping through the sub-soil, at any position may exert a force f, which may be tangential to streamline at any point. As the streamlines bend upward, the tangential force f will be having a vertical component f_1.

At that point, there will be a downward force W due to the submerged weight of the soil particle. Thus at that point there will be two forces on particle where one is upward vertical component of f and the other, is the submerged weight. It is evident that if the soil particle is not to be dislodged, then the submerged weight should be greater than upward vertical component of f. The upward vertical component force at any point is always proportional to the water pressure gradient dp/dx.

Thus for stability of the soil and for the prevention of erosion and piping, the seeping water when it emerges at the downstream side, at the exit position, the force f1 must be less than the submerged weight W. In other words the exit gradient at the downstream end should be safe.

If at the exit point at the downstream side, the exit gradient is such that force f_1 is equal to the submerged weight of soil particle, then this gradient is known as the Critical gradient. The Safe exit gradients = 0.2 to 0.25 of the critical exit gradient.

Values of the safe exit gradient can be taken as:

0.14 to 0.17 for fine sand.

0.17 to 0.20 for coarse sand.

0.20 to 0.25 for shingle.

For the standard form consisting of floor of length b and vertical cut-off of depth d, the exit gradient at its downstream end is given as:

Exit gradient GE = (H/d) x $\frac{}{\pi\sqrt{\lambda}}$

Where,

$$\lambda = \frac{1+\sqrt{1+\alpha^2}}{2}$$

Where,

$\alpha = b/d$

H = maximum seepage head.

2.8 Canal Falls

Irrigation canals are constructed with some permissible bed slopes so that there is no silting or scouring in the canal bed. But it is not always possible to run the canal at the desired bed slope throughout the alignment due to the fluctuating nature of country slope.

Generally, the slope of the natural ground surface is not uniform throughout the alignment. Sometimes, the ground surface may be steep and sometimes it may be very irregular with abrupt change of grade. In those cases, a vertical drop is provided to step down the canal bed and then it is continued with permissible slope until another step down is necessary. This is done to avoid unnecessary huge earth work in filling. Such vertical drops are known as canal falls or simply falls.

Necessity of Canal Falls

- When the slope of the ground suddenly changes to steeper slope, the permissible bed slope cannot be maintained. It requires excessive earthwork in filling to maintain the slope. In such a case falls are provided to avoid excessive earth work in filling.

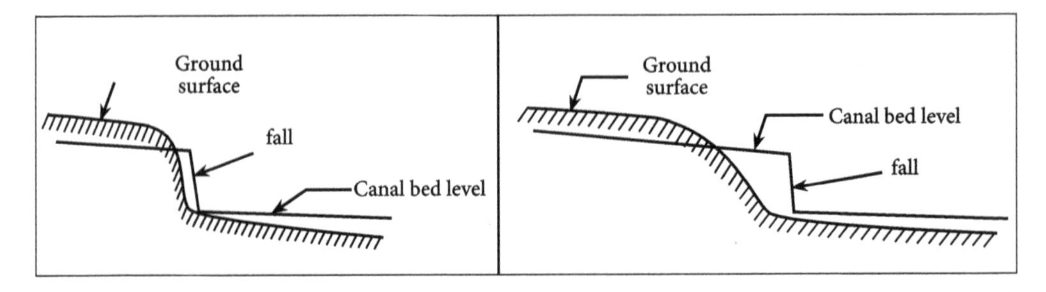

- When the slope of the ground is approximately uniform and the slope is greater than the permissible bed slope of canal, in that case also the canal falls are needed.

- In cross-drainage works, either when the difference between bed level of canal and that of drainage is small or when the F.S.L of the canal is above the bed level of drainage then the canal fall is necessary to carry the canal water below the stream or drainage.

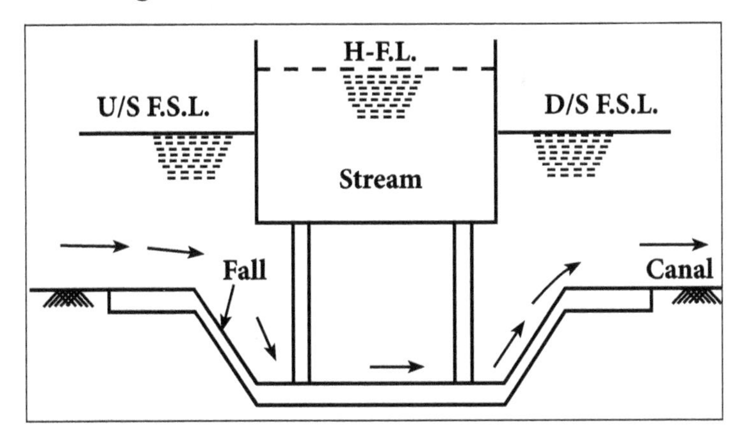

A canal fall is a hydraulic structure constructed across a canal to lower its water level. This is achieved by negotiating the change in bed elevation of the canal necessitated by the difference in ground slope and canal slope.

The necessity of a fall arises because the available ground slope usually exceeds the designed bed slope of a canal. Thus, an irrigation channel which is in cutting in its head reach soon meets a condition when it has to be entirely in filling. An irrigation channel in embankment has the disadvantages of:

- Higher construction and maintenance cost.
- Higher seepage and percolation losses.
- Adjacent area being flooded due to any possible breach in the embankment.
- Difficulties in irrigation operations. Hence, an irrigation channel should not be located on high embankments.

Therefore, falls are introduced at appropriate places to lower the supply level of an irrigation channel. The canal water immediately downstream and the fall structure possesses excessive kinetic energy which, if not dissipated, may scour the bed and banks of the canal downstream of the fall. This would also endanger the safety of the fall structure. Therefore, a canal fall is always provided with measures to dissipate surplus energy which, obviously, is the consequence of constructing the fall. The location of a fall is primarily influenced by the topography of the area and the desirability of combining a fall with other masonry structures such as bridges, regulators and so on.

In case of main canals, economy in the cost of excavation is to be considered. Besides, the relative economy of providing a large number of smaller falls (achieving balanced earth work and ease in construction) compared to that of a smaller number of larger falls (resulting in reduced construction cost and increased power production) is also worked out.

In case of channels which irrigate the command area directly, a fall should be provided before the bed of the channel comes into filling. The full supply level of a channel can be kept below the ground level for a distance of up to about 500 meters downstream of the fall as the command area in this reach can be irrigated by the channels off-taking from upstream of the fall.

Proper Location

The location of a fall in a canal depends upon the topography of the country through which the canal is passing. In case of the main canal, which does not directly irrigate any area, the site of a fall is determined by considerations of economy in 'cost of excavation and filling' versus 'cost of fall.

By providing a larger drop in one step, the quantity of unbalanced earth work increases, but at the same time, the number of fall reduces. An economy between these two factors has to be worked out before deciding the locations and extent of falls.

In case of branch canals and distributaries channels, the falls are located with consideration to commanded area. The procedure is to fix the FSL required at the head of the off-taking channels and outlets and mark them on the L-section of the canal. The FSL of the canal can then be marked, as to cover all the commanded points, thereby deciding suitable locations for falls in canal FSL and hence, in canal beds.

Types of Canal Falls

The following are the different types of canal falls that can be adopted according to the site condition:

Ogee Fall

In this fall, an ogee curve is provided for carrying canal water from higher level to the lower level. This fall is suitable when the natural ground surface suddenly changes to a steeper slope along the alignment of the canal.

- The fall consists of the concrete vertical wall and the concrete bed.
- Over the concrete bed the rubble masonry is present in the shape of ogee curve.
- The surface of the masonry is finished with rich cement mortar.
- The upstream and downstream side of the fall is protected by stone pitching with cement grouting.

- The design considerations of the ogee fall depends on the site condition.

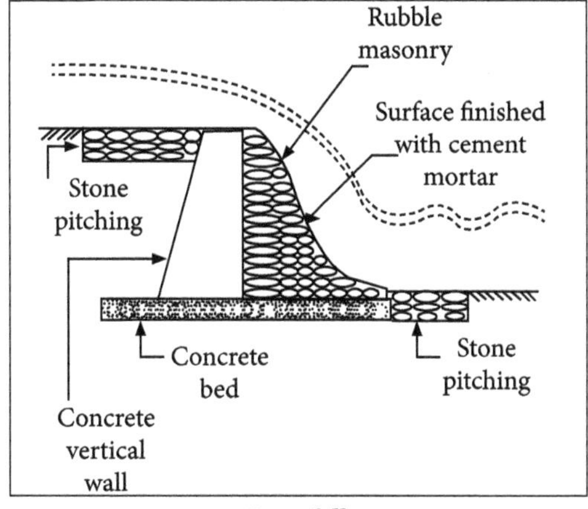

Ogee fall.

Rapid Fall

The rapid fall is apt when the slope of the natural ground surface is even and long. It has long sloping glacis with the longitudinal slope which varies from 1 in 10 to 1 in 20.

- Curtain walls were provided on the upstream and downstream side of the sloping glacis.
- The sloping bed is provided along with rubble masonry.
- The upstream and downstream side of the fall is also protected by rubble masonry.
- The masonry surface is finished with the rich cement mortar.

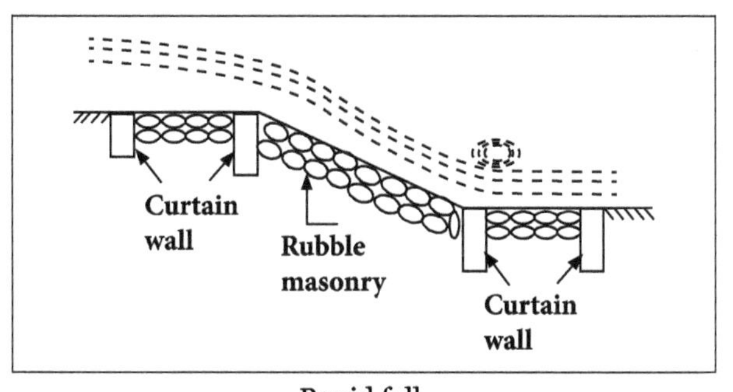

Rapid fall.

Stepped Fall

It consists of series of vertical drops in the form of steps. This fall is suitable in places where sloping ground is really long and requires long glacis to connect the higher bed level with lower bed level.

- This fall is practically a modification of the rapid fall.

- The sloping glacis is divided into a number of drops so that the flowing water will not cause any damage to the canal bed. Brick walls are provided at each of the drops.
- The bed of the canal within the fall is protected by rubble masonry with surface finishing by rich cement mortar.

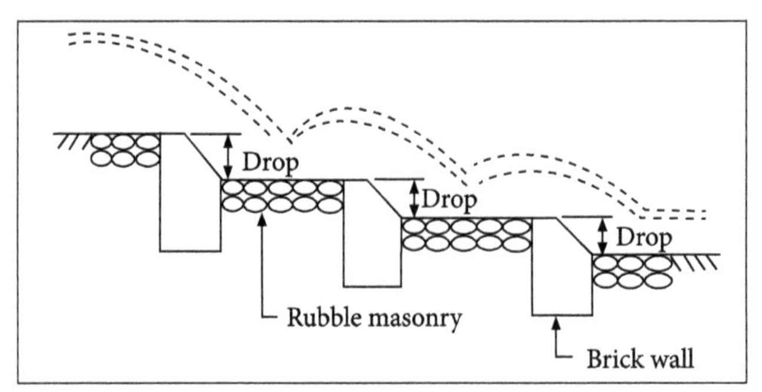

Stepped Fall.

Trapezoidal Notch Fall

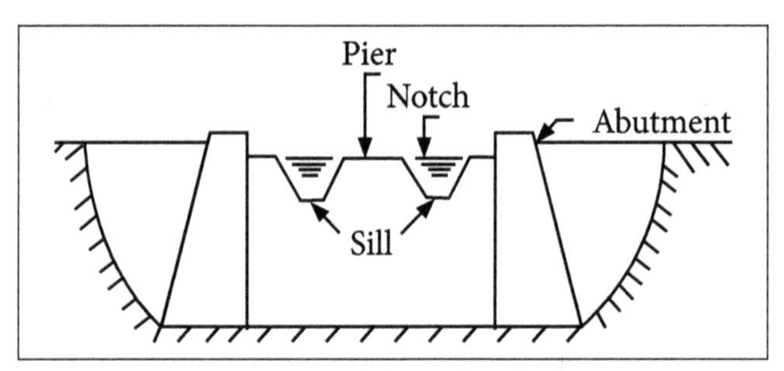

Trapezoidal Notch Fall.

In this fall a body wall is constructed across the canal. The body wall consists of number of trapezoidal notches between side piers and intermediate pier. The sills of the notches are kept at the upstream bed level of the canal.

- The body wall is constructed with the concrete or masonry.
- An impervious floor is then provided to resist scoring effect of falling water.
- The upstream and downstream side of fall is protected by means of stone pitching finished by cement grouting.
- The size and number of notches depends upon the full supply of the canal discharge.

Vertical Drop Fall

It consists of a vertical drop walls constructed with masonry work. The water flows over the crest of the wall. A water eastern is provided on the downstream side that acts as a water cushion in order to dissipate energy when the water falls.

- A concrete floor is present on the downstream side to control the scouring effect of the flowing water.
- Curtain walls are present on the upstream and downstream side.
- Stone pitching with cement grouting is provided on the upstream and downstream side of the fall so as to protect it from scouring.

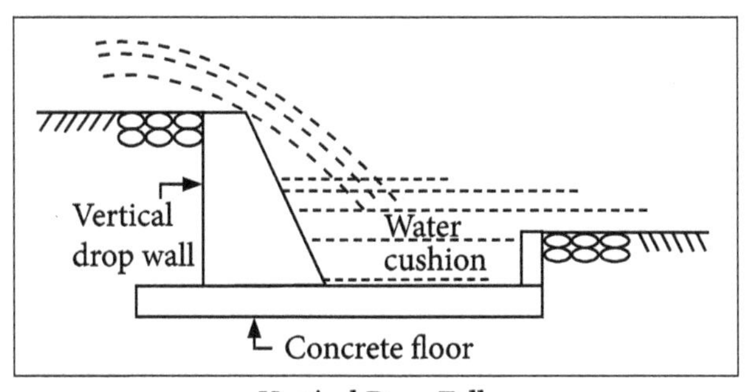

Vertical Drop Fall.

Glacis Fall

It consists of a straight sloping glacis provided along with the crest. A water cushion is provided on the downstream side to dissipate the energy of flowing water.

- The sloping glacis is constructed with cement concrete.
- Curtain walls and toe walls are provided on the upstream and the downstream side.
- The space between the toe walls and curtain walls is protected by means of stone pitching.
- This type of fall is suited for drops up to 1.5 m.

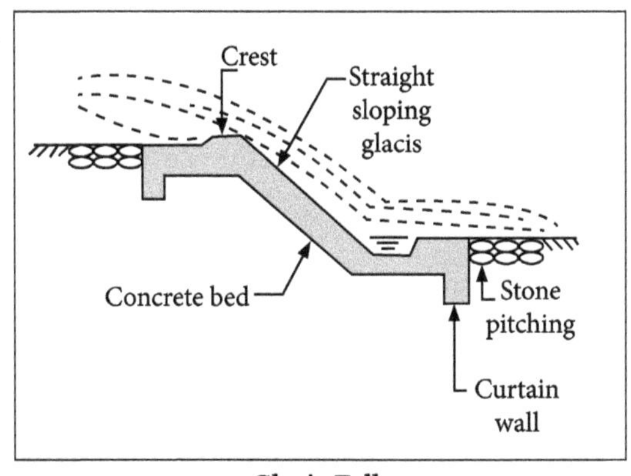

Glacis Fall.

(i) Montague Type Fall

In this type of fall, the straight sloping glacis is modified by giving a parabolic shape which is termed as the Montague profile. Taking "o" as the origin, the Montague profile is given by the equation,

$$X = \upsilon\sqrt{\frac{4y}{g} + Y}$$

Where, x = distance of point P from OX axis, Y = distance of point P from OY axis, υ = velocity of water at the crest, g = acceleration due to gravity.

The main body of the fall is constructed with cement concrete. Toe walls, curtain walls and bed protection by stone pitching are same as in the case of straight sloping glacis.

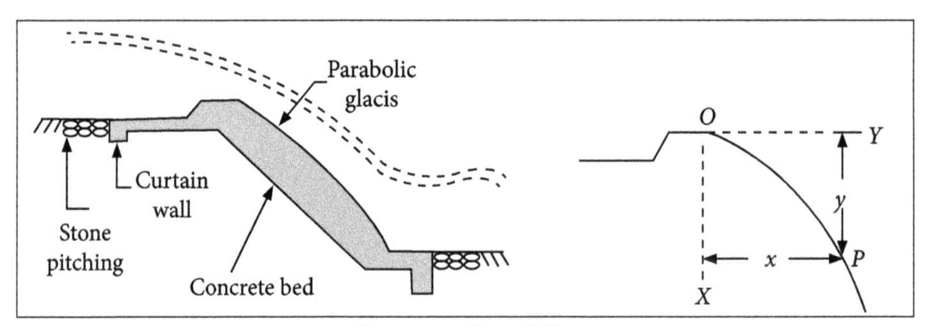

Montague Type Fall.

(ii) Inglis Type Fall

In this type of fall, the gracis is straight and sloping, but buffle walls are provided on the downstream floor to dissipate the energy of flowing water.

- The height of buffle depends on the head of water on the upstream side.
- The main body of the fall is constructed with the cement concrete.
- The toe walls and curtain walls are same as that of straight glacis.
- The protection works with stone pitching are also same. Sometimes, this fall is called as buffle fall.

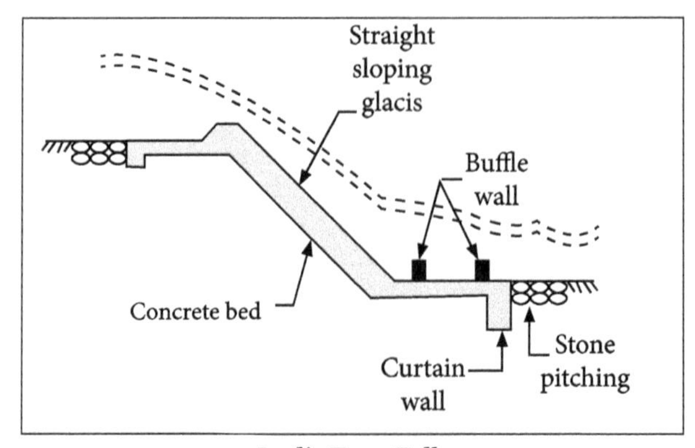

Inglis Type Fall.

2.8.1 Design and Detailing of One Type of Fall

Design Principles of Sarda Type Fall and Straight Glacis Fall

Such type of falls are constructed on Sarda canal in Uttar Pradesh. It is a fall with raised crest and with the vertical impact. The soils in Sarda command comprised sandy stratum overlain by sandy-clay on which the depth of cutting was to be kept minimum. This made it obligatory to provide the number of falls with small drops. In Sarda type falls (q) discharge intensity varied from 1.6 to 3.5 cumec/m and the drop varied from 0.6 to 2.5 m.

Crest Dimensions

This type of fall is not flumed.

For canal discharge 15 cumec,

Crest length of fall = Bed width of the canal.

For distributaries and minors,

Crest length of fall = Bed width + Depth of flow.

Body wall: When discharge of the canal is less than 14 m^3/sec then the section of body wall is kept rectangular (Figure (a)).

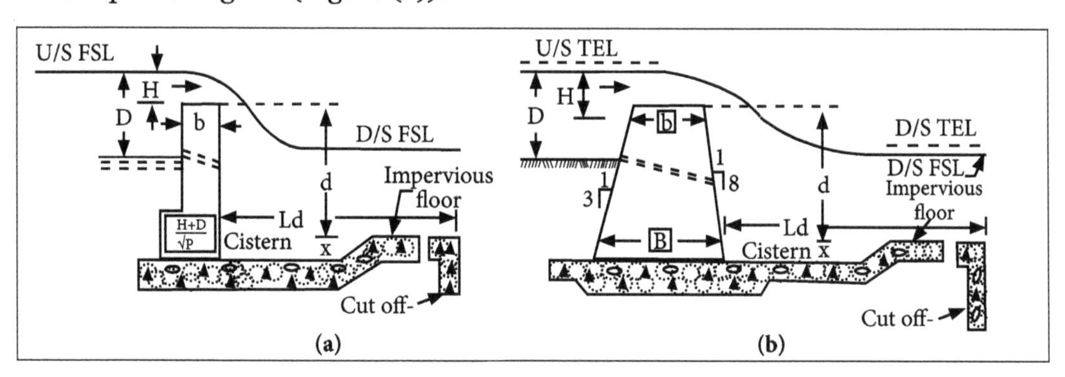

When the discharge of a canal is more than 14 m^3/sec then the section of the body wall is kept trapezoidal with upstream batter 1: 3 and downstream batter 1: 8.

For Rectangular Body Wall

Base width 'B' = $H + d/\sqrt{p}$.

Top width 'b' = $0.552\sqrt{d}$.

For trapezoidal body wall Top width $b = 0.522\sqrt{(H + d)}$.

The edges are rounded with the radius of 0.3 m.

Base width B is calculated by the batter given to u/s and d/s sides.

Where,

H is depth of water above the crest of the fall in meters.

d is the height of the crest above the downstream bed level in meters.

Discharge Over Crest

Under free fall condition the discharge formula used in this type of fall is:

Q = CLH {H/b} 1/6

Where L is length of crest in m and Q is discharge in cumec.

Value of C for trapezoidal crest is 2 and for rectangular crest 1.85.

For submerged flow conditions neglecting velocity of approach the discharge is given by the following formula,

$$Q = \frac{2}{3} Cd.L.\sqrt{2g}\,(H_L)^{3/2} + Cd.L.h_2\,\sqrt{2g.H_L}$$

Where,

Cd = 0.65.

H_L = Drop in water surface.

h_2 = Depth of d/s water level over top of crest.

Crest Level

The height of crest above the upstream bed level is fixed in a way that the depth of flow u/s of the fall is not affected. From the discharge formula mentioned above, as Q is known value of H can be calculated.

R.L of crest = F.S.L on the u/s – H.

The stability of body wall should be tested by the usual procedure when the drops exceeding 1.5 m are to be designed. In the body wall drain holes can be provided at the u/s bed level to dry out the canal during closures for maintenance, etc.

Cistern dimensions: Dimensions of the cistern can be fixed from the Bahadurabad Research Institute formula given, i.e.,

$LC = 5\sqrt{E.H_L}$

$$x = \frac{1}{4}(E.H_L)2/3$$

Total Length of Impervious Floor

For any hydraulic structure total length of impervious floor must be designed on the basis of the Bligh's theory for small structures and Khosla's theory for some other works. The maximum seepage head is experienced when on u/s water is up to the crest level of the fall and also when there is no flow on the d/s side. Referring the Figure maximum seepage head is given by 'd'.

Length of D/S Impervious Floor

The maximum length of the d/s impervious floor is given by the following relation.

$$Ld = 2D + 2.4 + H_L \text{ in meters.}$$

The balance of the impervious floor can be provided under body wall and on u/s.

Thickness of the Floor

The d/s floor must be made thick enough to resist the uplift pressures. However, minimum thickness of 0.3 to 0.6 m of concrete under 35 cm of brick masonry can be provided on the d/s. Masonry is not necessary on the u/s brick. The brick on the edge laid on the d/s impervious concrete floor provide the additional strength and affords easy repairs to the floor.

Cut-Off

A required depth of cut-off below the floor must be provided at the d/s end of the floor so as to provide safety against steep exit gradient. The depth of cut-off can range from 1 to 1.5 m.

Sometimes deeper cut-offs may be necessary to reduce horizontal floor length to satisfy Khosla's principle of exit gradient. For falls having 1 m and above head on the crest must be provided more cut-offs. Cut-off at u/s end of floor is provided and that may be smaller in depth.

Other Protective Works

Providing other accessories such as upstream wings, staggered blocks on the cistern floor, downstream wings, bed and side pitching is generally done on the basis of the thumb rules. For big structures, actual design calculations may be done. For general arrangement see Figure below.

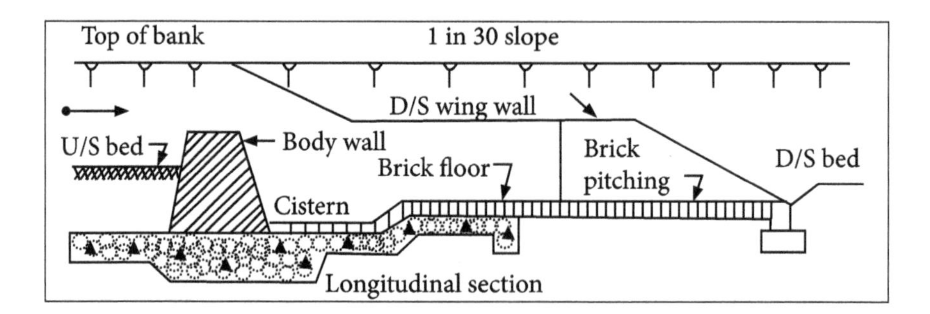

Upstream Wing Walls

For small falls up to 14 cumec upstream wings can be splayed at 1: 1. For higher discharges u/s wing walls are kept segmental with a radius equal to 6 H and continued thereafter tangentially merging to the banks. The wings can also be embedded into the bank for about 1 m.

Downstream Wing Walls

For the length of the cistern d/s wing walls are kept vertical from the crest. Thereafter they are wasped or flared to a slope of 1:1. An average splay of 1 in 3 for attaining required slope is given to the top of the wings. The wings can be taken deep into the banks.

Staggered Blocks

Staggered block of height dc must be provided at a distance of 1.0 dc to 1.5 dc from the d/s toe of crest for clear falls. In case of submerged falls the blocks may be provided at the end of the cistern. A row of staggered cubical blocks of height equal to 0.1 to 0.13 of depth of water is invariably provided at the end of the d/s impervious floor.

Bed and Side Pitching

The d/s bed pitching with bricks 20 cm thick over 10 cm ballast is provided horizontally for a length of 6 m. Thereafter for lengths up to 5 to 15 m and for falls varying from 0.75 to 1.5 m can be provided with down slope of 1 in 10.

The side pitching with bricks on edge with 1:1 slope is provided after the return-wing on the downstream. A toe wall should be provided between the bed pitching and the side pitching to provide a firm support to the latter.

Design Principles for Straight Glacis Fall: Crest Dimensions

Clear width of crest: Vertical falls must be full width falls, i.e., the width of the crest should be of similar bed width of the canal because increased intensity of the discharge due to fluming creates scour on the downstream. Unlike vertical falls the glacis falls may be flumed when combined with bridge so as to economize the cost. It is quite rational to select such (q) discharge per meter run of crest width with which the height of drop (H_L) available gives value of total energy on the d/s (Ef_2) equal to F.S. depth of the canal.

It does not need deep cistern on d/s and avoids construction difficulty particularly when the subsoil water level is high. The throat width can be rounded off to next half meter. The fluming thus calculated may not exceed its limit given below. It is subjected to the condition that overall width of fall crest is not more than the bed width of the canal on the downstream.

Drop (H_L)	**Permissible fluming ratio**
Upto 1 m	66%
1 to 3 m	75%
Above 3 m	85%

Crest level = u/s TEL – E

In case of full width falls and sometimes in flumed falls if the crest level works out unreasonably then high fluming may be done or increased. If already flumed so that crest is not higher than 0.4 –D_1, then above the u/s bed as otherwise it will increase afflux at low supplies and may cause alternate silting/scouring.

The value of E is calculated based on the discharge formula $Q = 1.84\ B_t \times E_3/2$.

Where,

B_t is clear width of crest. Therefore if n piers are provided between effective $B_t = (B_t - 0.2\ n\ H)$.

E is depth of crest below u/s TEL.

Length of crest (L_t) = 2/3 E.

The crest is joined to u/s and d/s canal bed with sloping glacis.

The u/s glacis (for non-meter falls) is given a slope of 1/2:1. The u/s crest end is kept curved with a radius of E/2.

The d/s glacis is given 2:1 slope and it is joined with the cistern d/s along the curve having radius equal to E.

Cistern Dimensions

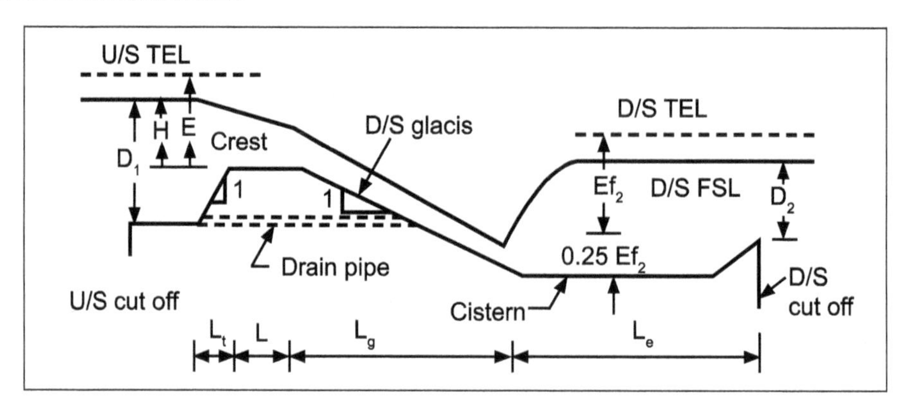

R.L. of cistern $= \mathrm{d/s\ TEL} - 1.25\ \mathrm{Ef}_2 = \mathrm{d/s\ FSL}\ - 1.25\ \mathrm{D}_2$

Length of cistern = 5 Ef_2 for good earthen bed or Ld = 6Ef_2 for erodible sandy soils.

The cistern is joined to the designed d/s bed with the up slope of 1 in 5 (1:5). This arrangement enables the formation of hydraulic jump on the sloping glacis.

Provisions of Cut-Offs

The cut-offs must be invariably provided at the upstream end of upstream glacis and also at the downstream end of the downstream cistern. The width of each curtain wall may be kept 0.4 m.

The depth can be as follows:

The Depth of u/s cut-off $= \mathrm{D}_1 / 3$

The Depth of d/s cut-off $= \mathrm{D}_2 / 2$

The minimum depth must be 0.5 m.

Total Length of the Impervious Floor

The total length of the floor should be with the depth of curtain walls as fixed earlier give permissible exit gradient. Khosla's curve for the exit gradient can be used for this purpose.

The length of the floor between u/s and d/s cut-offs so determined. If it appears to be excessive then the downstream cut-off can be further deepened accordingly to achieve adequate floor length.

It may be noted that total impervious floor length comprises:

- Length of the cistern.

- Horizontal length of the d/s glacis.
- Crest length along the axis of the canal.
- Horizontal length of the u/s glacis.

In some cases little length still remains to be provided as per the earlier calculations. It may be provided on the u/s side of the u/s glacis.

Thickness of the Floor

The Minimum thickness on the u/s can be from 0.3 to 0.6 m. Floor thickness in the glacis and the cistern must be sufficient to withstand uplift pressure safely.

U/s Approach and U/s Protection

- When the fall combines with the functions of a discharge meter as well, then the side and bed approaches to the crest are necessarily to be gradual and smooth, so as to avoid eddies and impact losses. It also reduces concentration of flow.
- In non-meter falls, the side walls can be splayed at an angle of 45° from upstream edge of the crest. The walls are carried straight into canal berm for a length of atleast 1 m.
- The bed approach can be by means of u/s glacis which has 1/2:1 slope and joining tangentially the u/s end of crest having radius equal to E/2.
- Protection of bed and sides by stone or dry brick pitching can be done for a length of (D_1 + 0.5) m. The bed pitching can be laid at the slope of 1 in 10.

D/s Expansion and D/s Protection

- On downstream, parallel and vertical walls are provided up to the toe of the glacis.
- The expansion afterwards must be gradual so that expanding flow adheres to the sides and scour due to the formation of back-rollers on sides is also prevented. A rectangular hyperbolic expansion is given by Mitra's equation for hyperbolic expansion can be adopted.
- If this expansion works out too long, side splay of about 1 in 5 can be adopted. For small falls to effect economy expansion with side splay of 1 in 3 is considered to be sufficient.
- Side walls in expansion may be flared out from vertical to 1: 1, if the earth fill behind is not problematic like black cotton soil. In those cases the side walls can be designed as vertical gravity walls.

- Side protection having 20 cm thickness of dry brick pitching for a length of $3D_2$ must be provided. It should rest on the toe wall of 1½ brick thick and of depth equal to $D_2/2$ subject to minimum of 0.5 m depth.
- The deflector wall of height $D_2/10$ above d/s bed can be provided at the downstream end of cistern. The minimum height should be 15 cm. Thickness of the deflector wall can be kept 0.4 m.
- With provision of deflector wall at the end of the floor d/s bed pitching beyond the floor is not required.

Friction Blocks as Energy Dissipators

Friction blocks are known to be the most effective energy dissipators. In case of flumed straight glacis falls four rows of friction blocks can be provided. They are staggered in the plan. The u/s edge of the first row of the friction block is located at the distance of 5 times the height of the blocks (5.h) from the toe of glacis.

The dimensions of the blocks may be as follows:

Let,

Height of the blocks = h, h = $D_1/8$.

Length of the block = 3h.

Width of the block = 2/3h.

Distance between rows = 2/3h.

When glacis is provided with baffle only two rows of friction blocks is essential up to 2 m fall. The u/s edge of the first row may be located at 1/3 length of d/s expansion from the end of the cistern floor.

Problems

Let us design a canal drop of the notch type for the following data:

Canal particulars	Upstream canal	Downstream canal
Bed width	8m	8m
Bed level	20.00	18.00
F.S.L	21.50	19.50
TBL Top of bank level	22.50	20.00

The ground at site is at 20.50, Good foundation level = 18.50. The side slopes for the canal may be taken as 1:1 in cutting and 1 ½: 1 in filling.

Let us also design the drop wall, the notches, solid apron. Draw to scale:

- ½ front and ½ sectional elevation.
- ½ plan at bottom and ½ at top.
- Cross section along the flow through one of the notches.

Solution:

Notches: Assume 3 notches and consider one notch for analysis.

The conditions of full supply and ½ supply will be as follows:

- Full supply discharge/notch = 8.0/3 m^3/s.
- Full supply depth(d) = 1.5m (on upstream side).
- Half supply discharge/notch= 4/3 m^3/s.
- Half supply depth (dc) = 2/3 x full supply depth=2/3 x 1.5= 1m.

Discharge through 1 trapezoidal notch is given by Q,

Type formula is,

$C_d = 0.7$, d_c= upstream depth

Substituting the 2 conditions, we get,

$$L + 0.6n = 0.7443$$

$$L + 0.4n = 0.6837$$

Where n = 0.303 and l = 0.56, say 0.6m

Also top width of notch = $L + nD_c$ (use full supply)

$$= 0.6 + 0.303 \times 1.5$$

$$= 1.05 = 1.1\text{mdc}$$

Notch Wall or Notch Pier

The top of the notch wall will be kept at upstream FSL= 21.50m.

The notch wall rests on the drop wall.

Thickness of notch wall = $d_c/2$ = 80cm.

The sides of the notch wall are kept vertical.

(Note: The length of the notch wall i.e. the distance from abutment to abutment measured at the top of the notch wall should be about 7/8 canal bed width).

Drop Wall

The figure shows part of the cross section EF, in which the screen is placed parallel to the flow, but the screen EF should pass through of the notches.

The top of the drop wall will be kept at the upstream bed level i.e. at 20.00.

The top width of the drop wall may be kept at 15-30cm more than that of the notch wall. Hence assume top thickness of 1m.

Note: The bottom width and R.L of bottom of the drop wall can be obtained only after designing the water cushion.

Water Cushion

The depth of water cushion x is obtained from $x + d_1 = 0.91dc$

$$X = 0.4m$$

Height of drop wall $H = Hl + x = 2.4m$

R.L of bottom of drop wall $= 20 - 2.4$

$$= 17.60$$

(Note: The apron is extended to form the foundation for the drop wall, we see that the top of the apron will also be at 17.60)

$$B = H + d_c/\sqrt{G}$$

$$= 2.6m \text{ where } G = 2.25$$

Provide symmetrical batter on upstream side and downstream sides for the drop wall.

Solid Apron

The top of the solid apron will be at 17.60. Thickness of apron $t = \frac{1}{2}\left(\sqrt{HI} + D_C = 1m\right)$

R_l of bottom of apron = 16.60

The minimum and maximum length of the apron from the toe A are,

$$L_{min} = 1.5 + 2 = 4.96m$$

$$L_{max} = 2dc + 2 = 6.50m$$

Assume L= 5.5m (or 5m or 6m)

Bed Pitching and Rivetment

(Note: Stone pitching provided on earth slopes is known as rivetment)

Length of upstream rivetment = 3dc = 5m.

Length of downstream rivetment = 4(d + H) = 14m.

The length of bed pitching will be kept rivetment = 7m.

Provide a 1:5 reverse slope to connect apron and downstream bed.

Abutment Wing Wall and Return

Note 1: For the drop wall, the abutment, wing wall, downstream return only, we extend the solid apron in order to form the foundation. Thus the bottom of the above walls will be at 17.60, It is in the top of the apron.

But, this level should satisfy the good foundations requirement (18.50) We see that in this case it is 17.60 well below good foundation level, however the apron is not extended below the upstream wing wall. For this wall we provide a separate strip foundation based on good foundation level as usual.

Note 2: The top of the bank is the highest level of the earth in the canal. The top of any wall in contact with the earth of the canal with respect to TBL. Thus on the upstream side the top of the wall must be at the upstream TBL of 22.50, similarly on the downstream side the top of the return which is in contact with the canal should be at the downstream TBL of 20.50, The abutment is in contact with the upstream TBL.

Note 3: Since the top of the upstream wing wall is kept level at the upstream TBL, we do not provide return on the upstream side. Also the downstream wing wall will slope down from 22.50(u/s TBL to d/s TBL) the top of the downstream return wall will be kept at the downstream TBL.

Note 4: The end of downstream wing wall will lie both at the end of the apron and the canal bed time. But on the upstream side there is no apron, hence the upstream wing wall will be splayed at 45. Also the upstream wing wall will be extended 50cm into the upstream top of the bank.

Similarly the downstream return will be extended by 50cm into the downstream top of the bank. This will automatically fix the length of upstream wing wall and downstream return.

Abutment

Top of the abutment will be at upstream TBL i.e. 22.50.

Its bottom will be at 17.60.

Bottom thickness = 0.4H=2m.

Side slope 1:8 batter on water side.

Length of the abutment at the bottom i.e. The distance at a will be kept = bottom width of the drop wall is 2.6m.

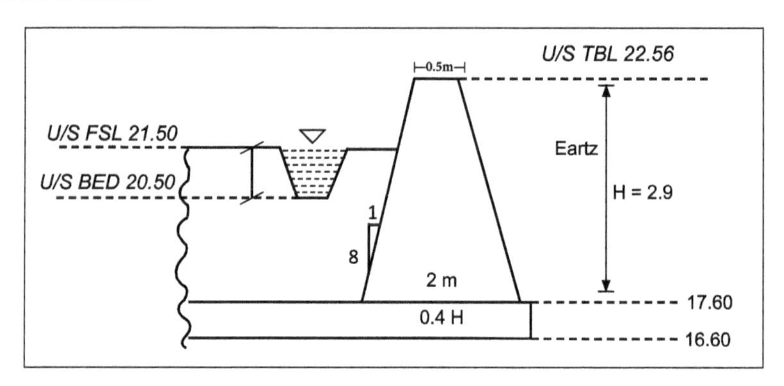

Downstream Wing Wall

Upstream TBL = 22.50 = top of abutment.

Downstream TBL = 20.50.

The top of the abutment is at 22.50.

Take of the downstream return at downstream TBL i.e. 20.50.

Thus top of upstream wing wall will slope down from 12.50 to 20.50.

Top of the downstream return will be level at 22.50. Hence we have the following sections.

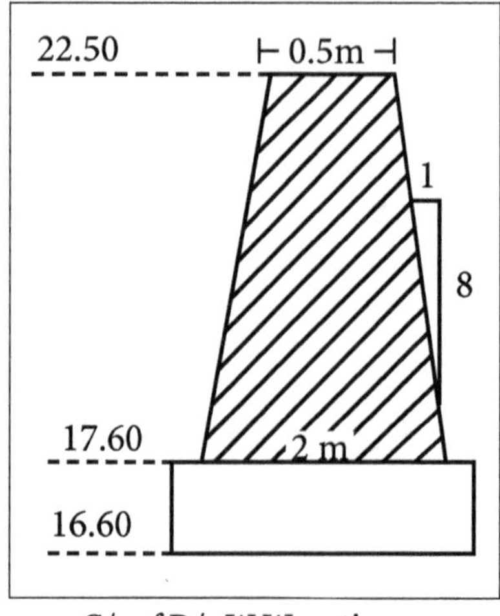

C/s of D/s W.W section-1.

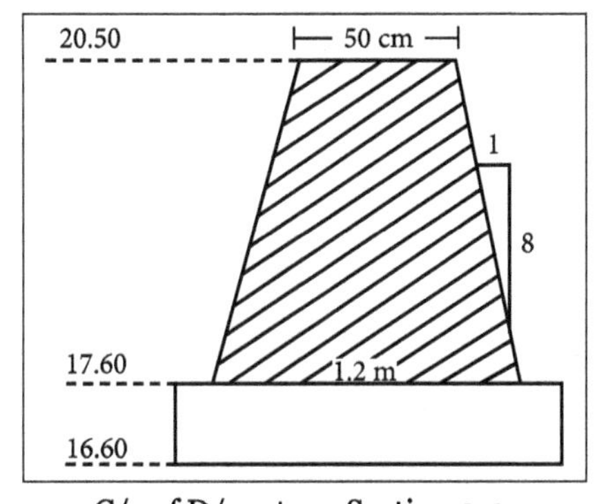

C/s of D/s return Section 2-2.

The downstream return will be extended by 50cm into the downstream top of the bank. The splay of the downstream will be automatically fixed in the drawing.

Upstream Wing Wall

The top of the upstream wing wall be kept level at the upstream TBL. i.e. At 22.50 Good foundation is at 18.50. Thus the bottom of the wall will be at the level as shown below. The upstream wing wall will be splayed by 45 with the flow direction. It can be extended by 50cm into the upstream bank as shown.

Note 1: The free board in the canal is the vertical distance between the FSL to the TBL. If not given it may be assumed between 75cm-1.5m, depending on the site of the canal.

Note 2: The Berm is the horizontal platform at Ground level but within the canal Cross section. Assume berm width 0.5m-3m.

Note 3: Side slopes of the canal may be assumed as cutting - 1:1 or 11/2:1 (H: V) Filling 11/2:1 or 2:1.

Note 4: The top of the bank is the highest level of the earth in the canal. This level should be above the FSL by a distance known as free board. The top width of the bank may be assumed from 1m-6m.

Note 5: Standard cross section of the canal. The canal may be partly in cutting and partly in filling or it may be fully in cutting as shown in figure.

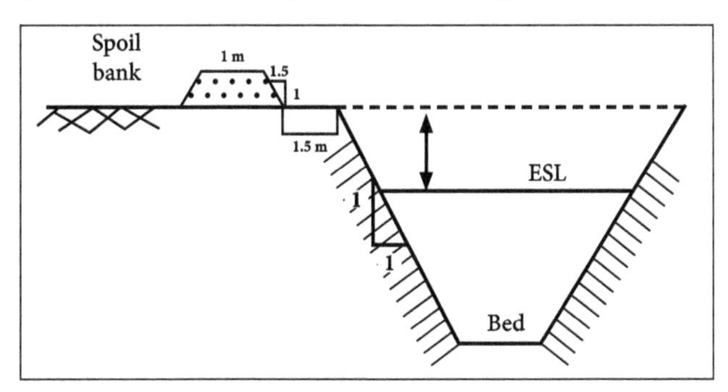

In this case the ground level gives sufficient free board. Hence cutting, filling is not required. However we provide small banks known as spoil banks as shown. In this case the TBL will be at Ground level. Large banks in filling are required if the Ground level is such that it cannot give sufficient fee board as shown below.

Note 6: In our case the cross section of the upstream and downstream canals will be as follows.

Upstream canal,

Bed level = 20.00

FSL = 21.50

GL = 20.50

We should see whether ground level can provide sufficient free board or not. In this case ground level cannot provide sufficient free board and hence we must provide fillings above ground level. If we assume free board of 1m above FSL the upstream TBL will be at 22.50, as given. Thus the section will be partly in cutting and partly in filling.

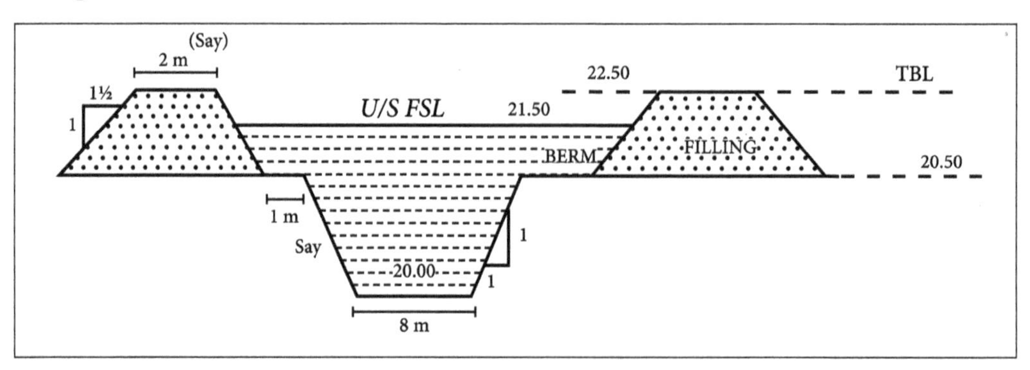

Downstream canal,

Bed level = 18.00

FSL =19.50

GL = 20.50

The ground level in this case is sufficient to provide which is 1m in this case. Hence the canal is in full cutting and ground level itself becomes the TBL on the downstream side. However provide small protection.

The figure shows the full sectional elevation AB, which is placed perpendicular to the flow.

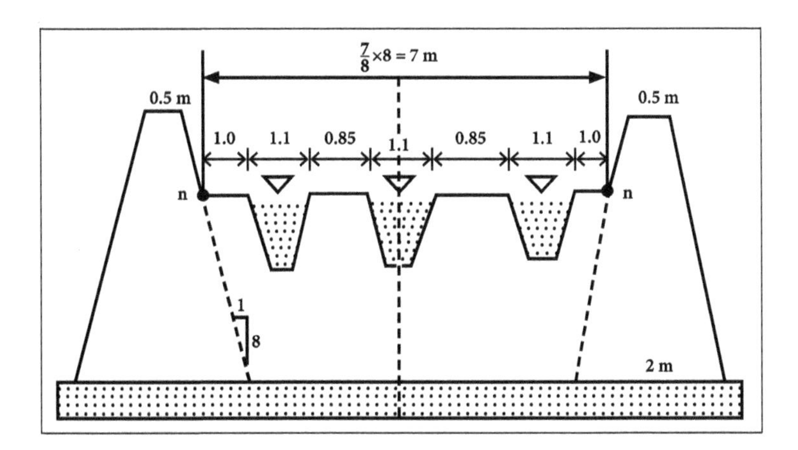

The length at the top of notch wall should be about 7/8 canal bed with which it becomes 7m in this case. The length of the notch wall between notches should not be less than dc/2. Further the length of the end notch walls should not be less than the intermediate walls. Also in this case the length of the notch at the top is 1.1m. Hence the length of various portions of notch wall can be assumed as shown.

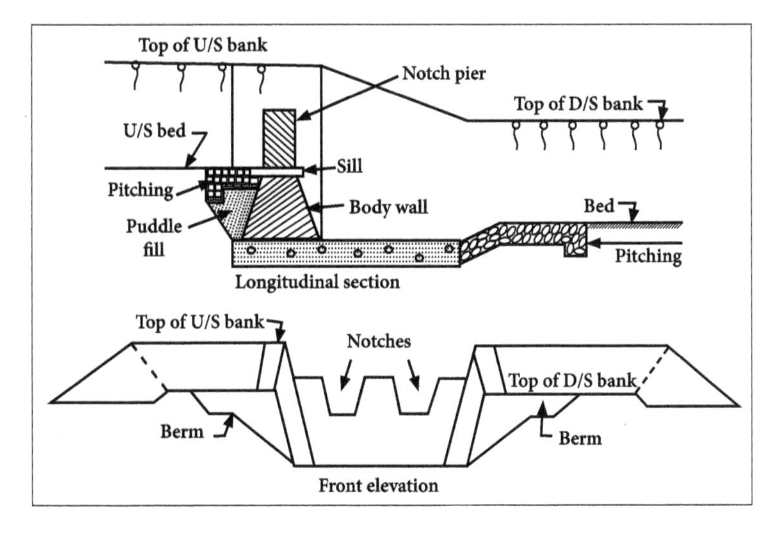

3

Gravity Dams, Earth Dams and Spillways

3.1 Gravity Dams

It is a masonry or concrete dam which resists the forces acting on it by its own weight. Its c/s is approximately triangular in shape.

Straight gravity dam: A gravity dam that is straight in plan.

Curved gravity dam: A gravity dam that is curved in plan. It resists the forces acting on it by combining the gravity action and arch action.

Solid gravity dam: Its body consists of a solid mass of the masonry or concrete.

Hollow gravity dam: It has hollow spaces within its body.

Most gravity dams are straight solid gravity dams.

Concrete Gravity Dams

Weight holds the dam in place.

Concrete gravity dam.

These dams are heavy and massive wall-like structures of concrete in which whole weight acts vertically downwards.

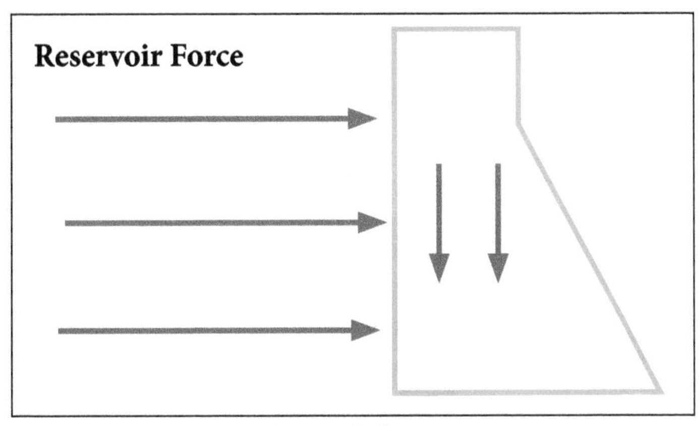

Reservoir force.

As the entire load is transmitted on the small area of foundation, such dams are constructed where rocks are competent and stable.

Bhakra Dam

It is the highest Concrete Gravity dam in Asia and is the second highest in the world. Bhakra Dam is across the river Sutlej in Himachal Pradesh. The construction of this project was started in the year 1948 and was completed in 1963. It is 740ft above the deepest foundation as straight concrete dam being more than three times the height of Qutub Minar.

Length at the top is 518.16m, width at the base is 190.5m and at the top is 9.14m Bhakra.

Bhakra Dam.

Gravity dam.

A gravity dam is any solid structure made of concrete or masonry built across a river to create a reservoir on its upstream. The section of gravity dam is approximately triangular in shape, with its apex at the top and the maximum width at the bottom.

The section is so proportioned that it resists the various forces acting on it by its own weight, usually consist of two sections, such as, the non-overflow section and the overflow section or spillway section are particularly suited across gorges with very steep side slopes and the earth dams might slip and are usually cheaper than the earth dams if suitable soils are not available for their construction where good foundations are available, the gravity dams shall be built up to any height.

It is the most permanent one and it requires only little maintenance. The most ancient gravity, dam on the record was built in Egypt more than 400 years B.C. of un-cemented masonry.

Basic Definitions

- Length of the dam: It is the distance from one abutment to other, measured along the axis of the dam at the level of top of the dam.
- Axis of the dam: It is the line of upstream edge of the top of the dam. The axis of dam in plan is called the base line of dam. And they are usually straight.
- Toe and Heel: The toe of the dam is the downstream edge of the base and the heel is the upstream edge of the base.
- Structural height of the dam: It is the difference in elevations of the top of the dam and the lowest point in the excavated foundation. It does not include the depth of special geological features of foundations like narrow fault zones below the foundation. In general, the height of the dam means its structural height.
- Hydraulic height of the dam: It is equal to the difference in elevations of highest controlled water surface on the upstream of the dam and the lowest point in the river bed.

- Maximum base width of the dam: It is the maximum horizontal distance between the heel and toe of the maximum section of the dam in the middle of valley.

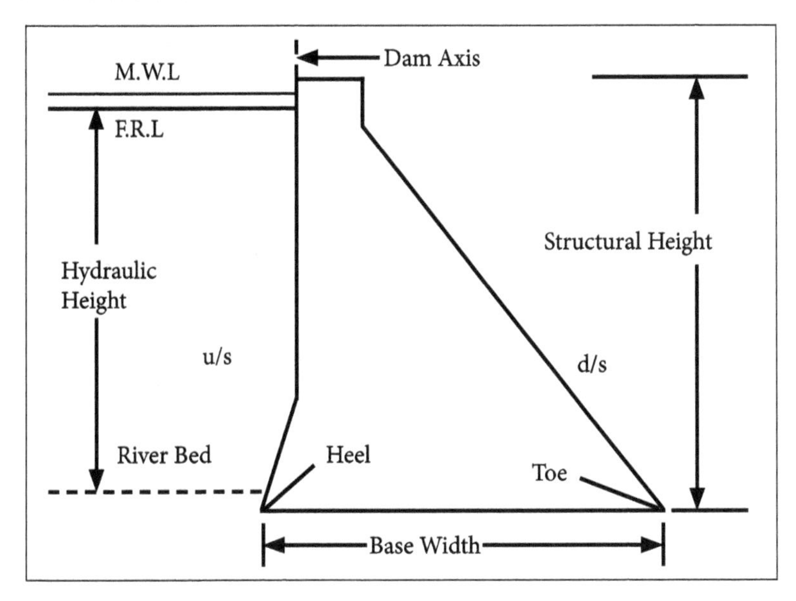

Typical Cross Section

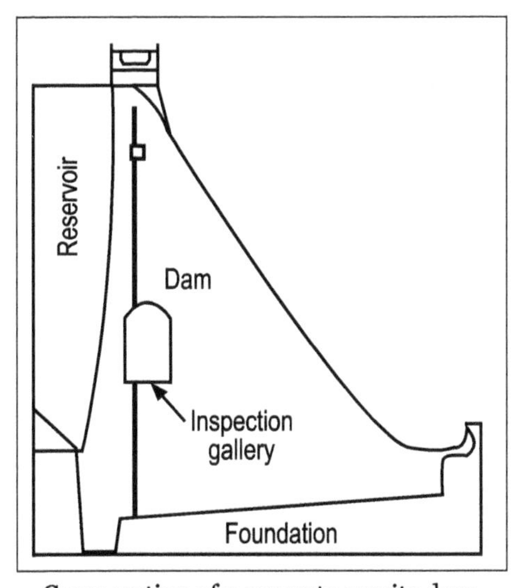

Cross section of a concrete gravity dam.

The different terms of importance are as follows:

- Maximum water level or full reservoir level: The maximum level to which the water rises during the worst flood is known as the maximum water level or full reservoir level.
- Minimum pool level: The lowest water surface elevation up to which the water in the reservoir can be used is called the minimum pool level.

- Normal pool level: It is the maximum elevation to which the reservoir water surface will rise during normal operating conditions.
- Useful and dead storage: The volume of water stored in the reservoir between the minimum pool level and normal pool level is called useful storage. The volume of water stored in the reservoir below the minimum pool level is known as dead storage.
- Free board: The margin between the maximum water level and top of the dam is known as free board. Free board must be provided to avoid the possibility of water spilling over the top of the clam due to wave action.
- Drainage gallery: A gallery provided near the foundation to drain off the water which seeps through the foundation and the body of darn is called the drainage gallery.

Various forces acting on gravity dam:

- Weight of the dam.
- Water pressure.
- Uplift pressure.
- Wave pressure.
- Earth and Silt pressure.
- Earthquake forces.
- Ice pressure.
- Wind pressure.
- Thermal loads.

These forces fall into two categories as:

- Forces such as weight of the dam and water pressure, which are directly calculable from the unit weights of the materials and properties of fluid pressures.
- Forces such as uplift, earthquake loads, silt pressure and ice pressure, which can only be assumed on the basis of assumption of varying degree of reliability.
- It is the second category of the forces so special care has to be taken and reliance placed on available data, experience and judgment.
- It is convenient to compute all the forces per unit length of the dam.

1. Weight of Dam

- Main stabilizing force in a gravity dam.
- Dead load = weight of concrete or masonry or both + weight of such appurtenances as piers, gates and bridges.
- Weight of the dam per unit length is equal to the product of the area of cross-section of the dam and the specific weight (or unit weight) of the material.
- Unit weight of concrete (24kN/m³) and masonry (23kN/m³) varies considerably depending upon the various materials that go to make them.
- For convenience, the cross-section of the dam is divided into simple geometrical shapes, such as rectangles and triangles, for the computation of weights. The areas and centroids of these shapes can be easily determined. Thus the weight components W_1, W_2, W_3 etc. can be found along with their lines of action. The total weight W of the dam acts at the C.G. of its section.

2. Water Pressure (Reservoir and Tail Water Loads)

Water pressure on the upstream face is the main destabilizing (or overturning) force acting on a gravity dam. Tail water pressure helps in the stability. Although the weight of water varies slightly with temperature, the variation is usually ignored. Unit Mass of water is taken as 1000kg/m³ and specific weight = $10kN/m_3$ instead of $9.81kN/m_3$.

The water pressure always acts normal to the face of dam. It is convenient to determine the components of the forces in the horizontal and vertical directions instead of the total force on the inclined surface directly. The water pressure intensity p (kN/m²) varies linearly with the depth of the water measured below the free surface y (m) and is expressed as $P = \gamma_w y$.

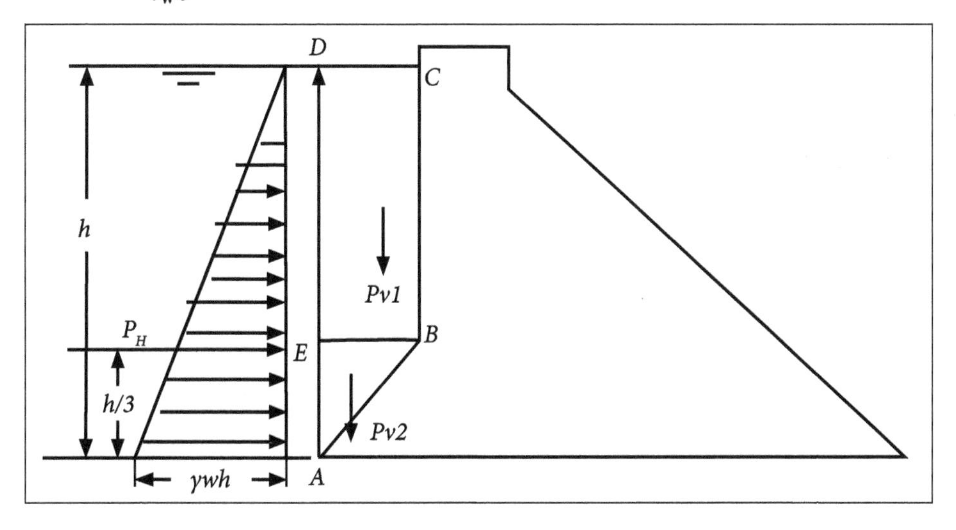

Water Pressure (Reservoir and Tail Water Loads) U/s Face Vertical

When the upstream face of the dam is vertical, the water pressure diagram is triangular in shape with a pressure intensity of $\gamma_w h$ at the base, where h is the depth of water. The total water pressure per unit length is horizontal, It acts horizontally at a height of h/3 above the base of the dam.

Water Pressure (Reservoir and Tail Water Loads) U/s Face Inclined

When the upstream face ABC is either inclined or partly vertical and partly inclined, the force due to water pressure can be calculated in terms of the horizontal component P_H and the vertical component P_V. The horizontal component acts horizontal at a height of (h/3) above the base.

The vertical component P_V of water pressure per unit length is equal to the weight of the water in the prism ABCD per unit length. For convenience the weight of water is found in two parts P_{V1} and P_{V2} by dividing the trapezium ABCD into a rectangle BCDE and a triangle ABE.

Thus the vertical component $PV = PV_1 + PV_2$ = weight of water in BCDE + weight of water in ABE. The lines of action of P_{V1} and P_{V2} will pass through the respective centroids of the rectangle and triangle.

3. Uplift Pressure

- Water has a tendency to seep through the pores and fissures of the material in the body of the dam and foundation material and through the joints between the body of the dam and its foundation at the base. The seeping water exerts pressure.
- The uplift pressure is defined as the upward pressure of water as it flows or seeps through the body of dam or its foundation.
- A portion of the weight of the dam will be supported on the upward pressure of water, hence net foundation reaction due to vertical force will reduce.
- The area over which the uplift pressure acts has been a question of investigation from the early part of this century.
- One school of thought recommends that a value one-third to two-thirds of the area should be considered as effective over which the uplift acts.

Code of Indian Standards (IS : 6512-1984)

- There are two constituent elements in uplift pressure: the area factor or the percentage of area on which uplift acts and the intensity factor or the ratio which the actual intensity of uplift pressure bears to the intensity gradient extending from head water to tail water at various points.

- The total area should be considered as effective to account for uplift.
- The pressure gradient shall then be extending linearly to heads corresponding to reservoir level and tail water level.

In case of drain holes: the uplift pressure at the line of drains exceeds the tail water pressure by one-third the differential between the reservoir and tail water heads. The pressure gradient shall then be extended linearly to heads corresponding to reservoir level and tail water level.

In case of a crack: The uplift is assumed to be the reservoir pressure from the u/s face to the end of the crack and from there to vary linearly to the tail water or drain pressure.

In absence of line of drains and for the extreme loading conditions F and G, the uplift shall be taken as varying linearly from the appropriate reservoir water pressure at the u/s face to the appropriate tail water pressure at the d/s face.

Uplift pressures are not affected by earthquakes.

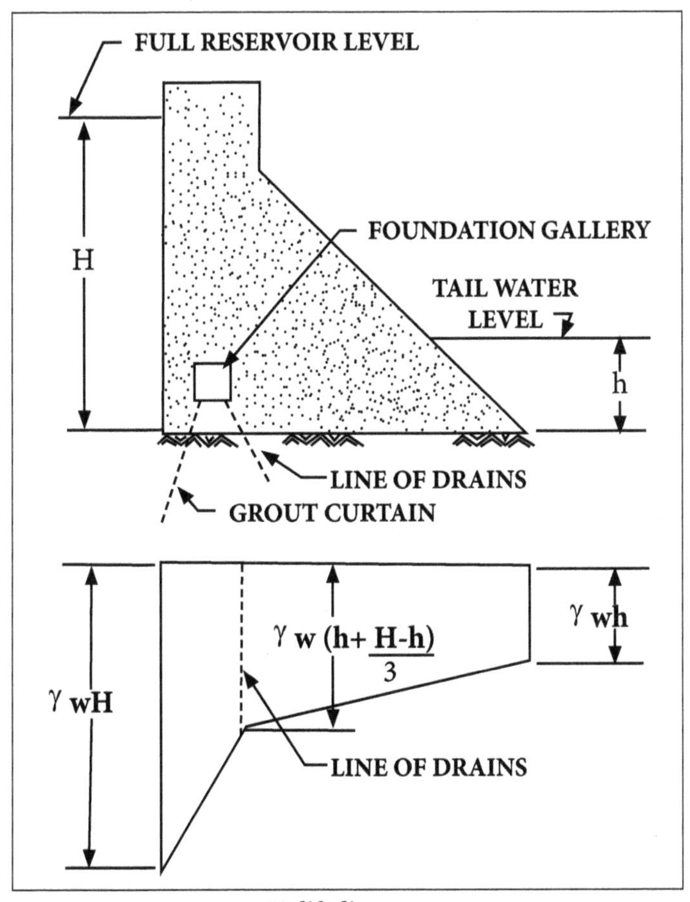

Uplift diagram.

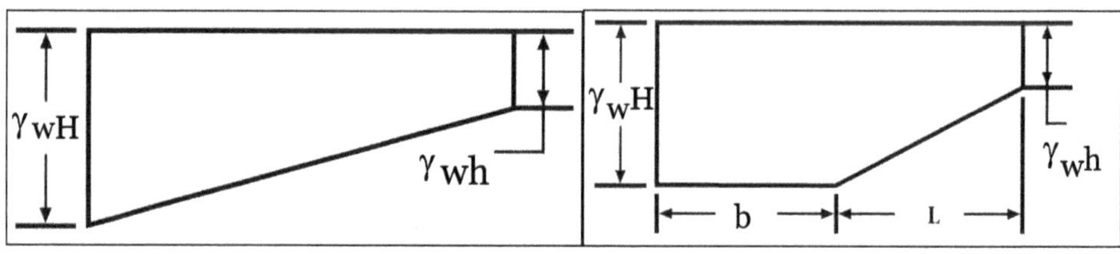

(a) Uplift pressure and (b) Final uplift pressure.

Diagram before cracking diagram for the cracked Section.

4. Earth and Silt Pressure

Gravity dams area unit is subjected to earth pressures on the downstream and upstream faces wherever the muse trench is to be backfilled. Except within the abutment sections in specific cases, earth pressures have sometimes a minor impact on the steadiness of the structure and will be unnoticed.

Silt is treated as a saturated cohesion less soil having full uplift and whose price of internal friction is not materially modified on account of submerging. IS code recommends that,

- Horizontal silt and water pressure is assumed to cherish that of a fluid with a mass of 1360 kg/m^3.
- Vertical silt and water pressure is set as if silt and water along have a density of 1925kg/m^3.

5. Ice Pressure

Ice expands and contracts with changes in temperature. In a reservoir utterly frozen over, the air temperature within the level of the reservoir water might cause the gap from cracks that afterward fill with water and frozen solid. Once succeeding rise in temperature happens, the ice expands and if restrained, it exerts pressure on the dam.

Good analytical procedures exist for computing ice pressures, however the accuracy of results relies upon sure physical information that haven't been adequately determined. Ice pressure is also provided for the speed of 250kPa to the face of dam over the anticipated space of contact of ice with the face of dam.

6. Wind Pressure

Wind pressure will exist however is rarely a major consideration in the planning of a dam. Wind masses might, therefore, be ignored.

7. Thermal Load

The cyclic variation of air temperature and the radiation on the downstream facet and also the reservoir temperature on the upstream facet have an effect on the stresses within the dam. Even the deflection of the dam is most within the morning and it goes on reducing to a minimum within the evening.

Measures for temperature management of concrete in solid gravity dams are adopted throughout construction. Thermal aren't important in gravity dams and will be unnoticed.

3.2 Combination of Forces for Design

Case 1: Reservoir Full Case

When reservoir is full, the major forces acting are weight of the dam, external water pressure, uplift pressure and earthquake forces in serious seismic zones. The minor forces are: Silt Pressure, Ice Pressure and wave Pressure. For the most conservative designs and from purely theoretical point of view one can say that a situation may arise when all the forces may act together.

But such a situation will never arise, hence, all the forces are not taken generally together. U.S.B.R. has classified the normal load Combinations and extreme load combinations as given below:

1. Normal Load Combinations

- Water pressure up to normal pool level, normal uplift, silt pressure and ice pressure. This class of loading is taken when ice force is serious.
- Water pressure up to normal pool level, normal uplift, earthquake forces and silt Pressure.
- Water Pressure up to maximum reservoir level (maximum pool level), normal uplift and Silt Pressure.

2. Extreme Load Combinations

- Water pressure due to maximum pool level, extreme uplift pressure without any reduction due to drainage and silt pressure.

Case 2: Reservoir Empty Case

- Empty reservoir without earthquake forces to be computed for determining bending diagrams, etc. for reinforcement design, for grouting studies or other purposes.
- Empty reservoir with a horizontal earthquake force produced towards the upstream has to be checked for non- development of tension at toe.

3.3 Modes of Failure and Criteria for Structural Stability

Following are the modes of failure of a gravity dam:

- Overturning.
- Sliding.
- Compression or Crushing.
- Tension.

1. Overturning

The overturning of the dam section takes place when the resultant force at any section cuts the base of the dam downstream of the toe. In that case the resultant moment at the toe becomes clockwise (or-ve). On the other hand, if the resultant cut the base within the body of the dam, there will be no overturning.

For stability requirements, the dam must be safe against overturning. The factor of safety against overturning is defined as the ratio of the right moment (+ve) to the overturning moments, i.e.,

$$\text{F.S} = \frac{\sum \text{Righting moments}}{\sum \text{Overturning moments}} = \frac{\sum M_R}{\sum M_o} \qquad ...(1)$$

The factor of safety against overturning should not be less than 1.5.

2. Sliding

A dam will fall in sliding at its base or at any other level, if the horizontal forces causing sliding are more than the resistance available to it at that level. The resistance against sliding may be due to friction alone or due to friction and shear strength of the joint.

Shear strength develops at the base if benched foundations are provided and at other joints if the joints are carefully laid so that a good bond develops. Shear strength also comes into play because of the interlocking of stone in masonry dams.

If the shear strength is not taken into account, the factor of safety is known as factor of safety against sliding. The factor of safety against sliding is defined as the ratio of actual coefficient of static friction (μ) on the horizontal joint to the sliding friction. The sliding factor is the minimum coefficient of friction required to prevent sliding.

If ΣH= sum of the horizontal forces causing the sliding and ΣV is the net vertical forces, the sliding factor ($\tan \theta$) is given by,

$$\text{S.F.} = \tan\theta = \frac{\sum H}{\sum V} \qquad ...(2)$$

And factor of safety against sliding (F.S.S) is,

$$S.F.F = \frac{\mu}{\tan\theta} = \frac{\mu\sum V}{\sum H} \qquad ...(3)$$

The coefficient of friction μ varies from 0.65 to 0.75. The factor of safety against sliding should be greater than the value given in Table (IS 6512-1972). It is considered that a low gravity dam should be safe against sliding considering friction alone. However, in large dams, shear strength of the joint should also be considered for an economical design. The factor of safety in that case is commonly known as the shear friction factor S.F.F. and is defined by the equation,

$$S.F.F = \frac{\mu\sum(V) + A.c}{\sum H} = \frac{\mu\sum(V) + V.c}{\sum H} \qquad ...(4)$$

Where,

c = Average cohesion or shear strength of the joint the value of which varies, from 1300kN/m² to 4500kN/m², for good rock to 650 to 1300kN/m² for concrete.

b = Width of the joint or section.

A = Area of joint = b x 1 for unit length of the dam.

Table: Factor of Safety against Sliding and Shear Friction Factor (IS : 6512 - 1972)

S. No.	Loading Condition	F.S. against sliding (F.S.S)	S.F.F
1	A,B,C	2.0	4.0
2	D,E	1.5	3.0
3	F,G	1.2	1.5

Recommendations of IS 6512-1984

The above criterion (i.e. Equation (4) and Table) given in IS 6512-1972 was completely changed in IS 6512-1984. According to the revised version, the factor of safety against sliding is calculated on the basis of partial factor of safety in respect of friction (F_Φ) and partial factor of safety in respect of cohesion or shear (F_c) by using the following equation:

$$F = \frac{\frac{\mu\sum(V)}{F_\varphi} + \frac{cA}{F_c}}{\sum H} = \frac{\frac{\mu\sum V}{F_\varphi} + \frac{cb}{F_c}}{\sum H} \qquad ...(5)$$

Where,

F = Factor of safety against sliding.

A = Area of the joint = b x 1 for unit length of the dam.

The values of F_Φ and F_c are given in Table below.

Table: Values of Partial Safety Factors F_Φ and F_c (IS 6512-1984):

S. No.	Load combination	F_Φ	F_c		
			For dams and the contact plane with foundation	For foundation	
				Thoroughly investigated	others
(i)	A,B,C	1.5	3.6	4.0	4.5
(ii)	D,E	1.2	2.4	2.7	3.0
(iii)	F,G	1.0	1.2	1.35	1.5

The factor of safety against sliding, computed from Equation (5), should not be less than (1).

3. Compression or Crashing

In order to calculate the normal stress distribution at the base or at any section, let H be the total horizontal force, V be the total vertical force and R be the resultant forces cutting the base at an eccentricity e from the centre of the base of width b (Figure(1)).

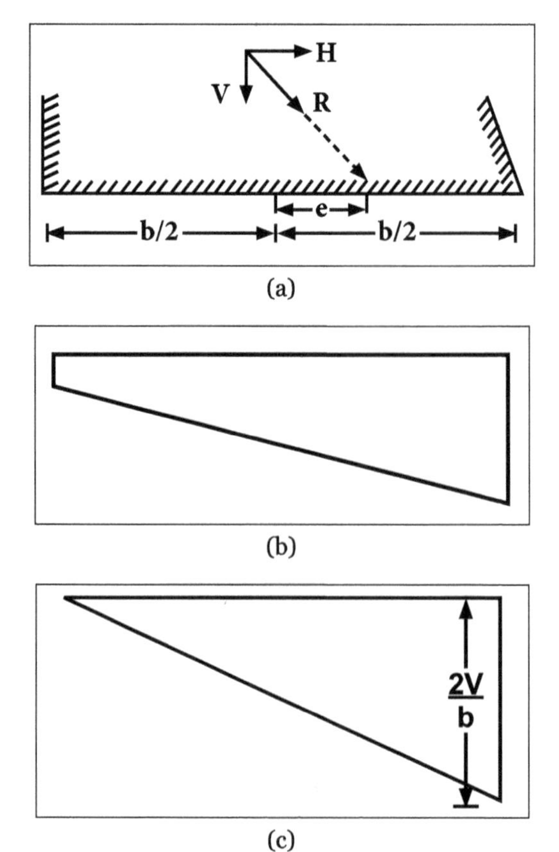

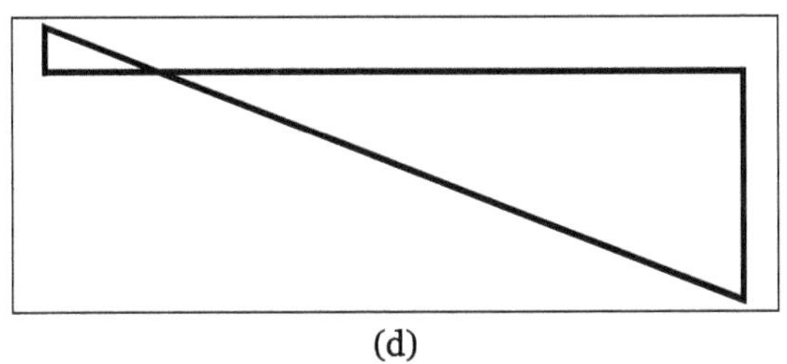

(d)
(1) Normal stress distribution at base.

The normal stress at any point on the base will be the sum of the direct stress and the bending stress. Thus, direct stress,

$$= \frac{V}{b \times 1}$$

Bending stress,

$$\pm \frac{My}{I} = \pm \frac{V.e}{\frac{1}{6}b^2} = \frac{6V.e}{b^2}$$

Hence the total normal stress P_n is given by b,

$$P_n = \frac{V}{b}\left(1 \pm \frac{6e}{b}\right) \qquad ...(6)$$

The positive sign will be used for calculating normal stress at the toe, since the bending stress will be compressive there and negative sign will be used for calculating normal stress at the heel.

Thus, the normal stress at the toe is,

$$(P_n)toe = \frac{V}{b}\left(1 + \frac{6e}{b}\right) \qquad ...(7)$$

The normal stress at the heel is,

$$(P_n)heel = \frac{V}{b}\left(1 - \frac{6e}{b}\right) \qquad ...(8)$$

Figure (1b) shows the normal stress distribution for a general case when the pressure at both toe and heel are compressive. Evidently, the maximum compressive stress occurs at the toe and for safety, this should not be greater than the allowable compressive stress *f* for the foundation material. Hence, from strength point of view,

$$\frac{V}{b}\left(1 + \frac{6e}{b}\right) \leq f \qquad ...(9)$$

When the eccentricity e is equal to b/6, we get,

$$(P_n)\text{toe} = \frac{V}{b}\left(1+\frac{6}{b}\times\frac{b}{6}\right) = \frac{2V}{b} \qquad \text{...(10)}$$

The corresponding stress at the heel in that circumstance will evidently be zero. Figure (1c) shows the pressure distribution for this case.

4. Tension

From equation (8) the normal stress at the heel is,

$$(P_n)\text{heel} = \frac{V}{b}\left(1-\frac{6e}{b}\right)$$

It is evident that if e >b/6 the normal stress at the heel will be -ve or tensile. No tension should be permitted at any point of the dam under any circumstance for moderately high dams.

For no tension to develop the eccentricity should be less than b/6. In other words, the resultant should always lie within the middle third. However, in case of extra high dams, 230 to 260 m, small tension within the permissible limit is generally permitted for comparatively small periods of loading such as heavy flood or earthquake.

Effect of Tension Cracks

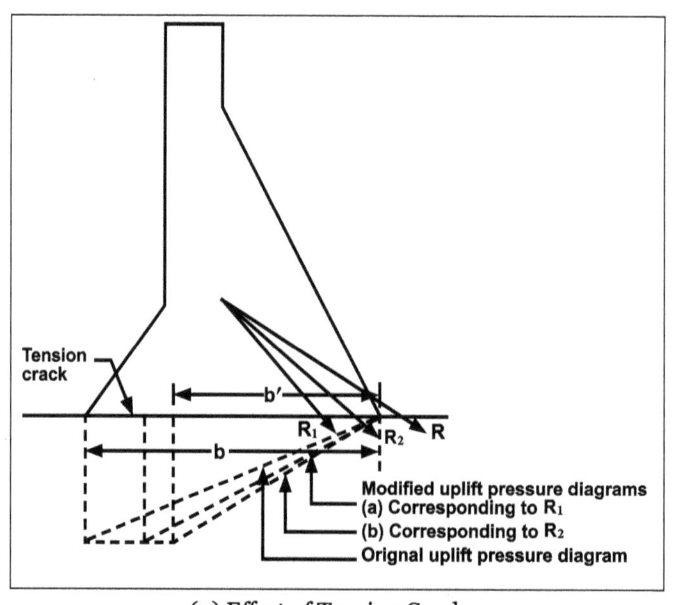

(2) Effect of Tension Cracks.

Since concrete cannot resist the tension, a crack develops at the heel, which modifies the uplift pressure diagram, as illustrated in Figure (2). Due to tension crack, the uplift pressure increases in magnitude and net downward vertical force or the stabilizing force reduces.

The resultant force thereby gets further shifted towards the toe and this leads to further lengthening of the crack. The base width thus goes on reducing and the compressive stresses on toe goes on increasing, till the toe fails in compression.

3.4 High and Low Gravity Dam

Normal stress in a dam is given by,

$$P_n = \frac{V}{b}\left(1 \pm \frac{6e}{b}\right)$$

Normal stress in a dam is given by,

Where,

$V = \Sigma$ (w-U) and e = b/6.

For stress reservoir full condition, p_n, at the toe is,

$$P_n = \frac{\sum(W-U)}{b}\left(1+\frac{6e}{b}\right)$$

$$P_n = \frac{\frac{1}{2}bHsw - \frac{1}{2}CbHw}{b}\left[1+\frac{6}{b}\times\frac{b}{6}\right]$$

or,

$p_n = wH(s - C)$ ...(1)

And at the heel,

$$P_n = \frac{\sum(w-U)}{b}\left[1-\frac{6}{b}\times\frac{b}{6}\right] = 0$$

Principal stress at the toe when no water downstream derived already is given by,

$$\sigma = p_n \sec^2 \Phi = p_n\left(1+\tan^2\Phi\right)$$

$$\sigma = wH(s-C)\left[1+\left(\frac{b}{H}\right)^2\right]$$

Substituting p_n from Equation (1),

$$\sigma = wH(s-C)\left[1+\frac{b^2}{H^2}\right] \qquad ...(2)$$

But from equation of elementary profile,

$$\frac{b}{H} = \frac{1}{\sqrt{s-c}}$$

$$\frac{b^2}{H^2} = \frac{1}{s-C}$$

Substituting b^2/H^2 in Equation (2) now,

$$\sigma = wH(s-C)\left(1+\frac{1}{s-C}\right)$$

$$\sigma = wH(s-C)\frac{s-C+1}{(s-C)}$$

$$\sigma = wH(s-C+1) \qquad ...(3)$$

Shear stress at toe when no water downstream is given by,

$$\lambda = P_n \tan\Phi$$

$$\lambda = wH(s-C)\frac{b}{H}$$

Replacing b by,

$$b = \frac{H}{\sqrt{s-C}}$$

$$\lambda = wH(s-C)\frac{H}{H\sqrt{s-C}}$$

$$\therefore \lambda = wH\sqrt{s-C} \qquad ...(4)$$

Principal stress should not exceed the safe allowable stress *f* of the material,

$$f = \sigma = wH(s-C+1)$$

$$H = \frac{f}{w(s-C+1)} \qquad ...(5)$$

For finding the limiting height H, it is not generally considered uplift force, hence, C = 0

$$H = \frac{f}{w(s+1)} \qquad ...(6)$$

If the height H is given by Equation (6), maximum compressive stress will exceed the permissible stress which is not desirable.

Equation (6) defines the limiting height and distinction between high and low gravity dam. A dam is called a low gravity dam whose height is less than the height given by Equation (6), so that maximum compressive stress is not greater than the allowable compressive stress of concrete.

Assume the specific gravity of concrete = 2.4

Safe allowable stress of concrete is 2.943 x 10_6 N/m²

∴ Limiting height of gravity dam given by Equation (6) is,

$$H = \frac{2.943 \times 10^6}{9.81 \times 1000(2.4+1)} m$$

$$H \approx 88 \text{ m}$$

Thus, for concrete gravity dam, limiting height is around 88 m.

Thus, a gravity dam is low if it is less than the height given by Equation (6) (or less than approximately 88 m) and the dam is high if it is more than H given by Equation (6) (or more than approximately 88 m).

Thus, the profiles of low and high gravity dam is shown in below Figure.

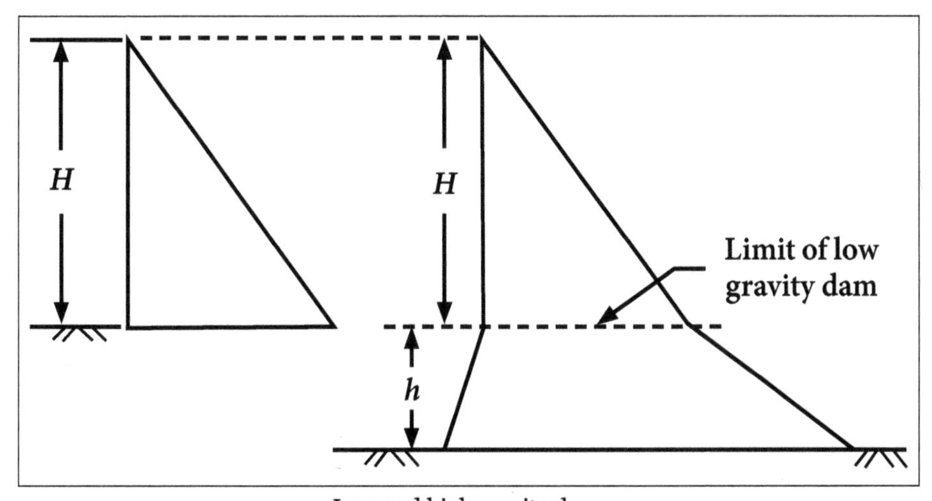

Low and high gravity dam.

Design of High Dam

Multiple Step Method of Design of Gravity Dam

For the economic design of gravity dam, the section of the dam is divided into various zones. Each zone is designed in such a way that all the requirements of stability are satisfied.

Figure (1) shows typical seven zones in a non-overflow gravity dam.

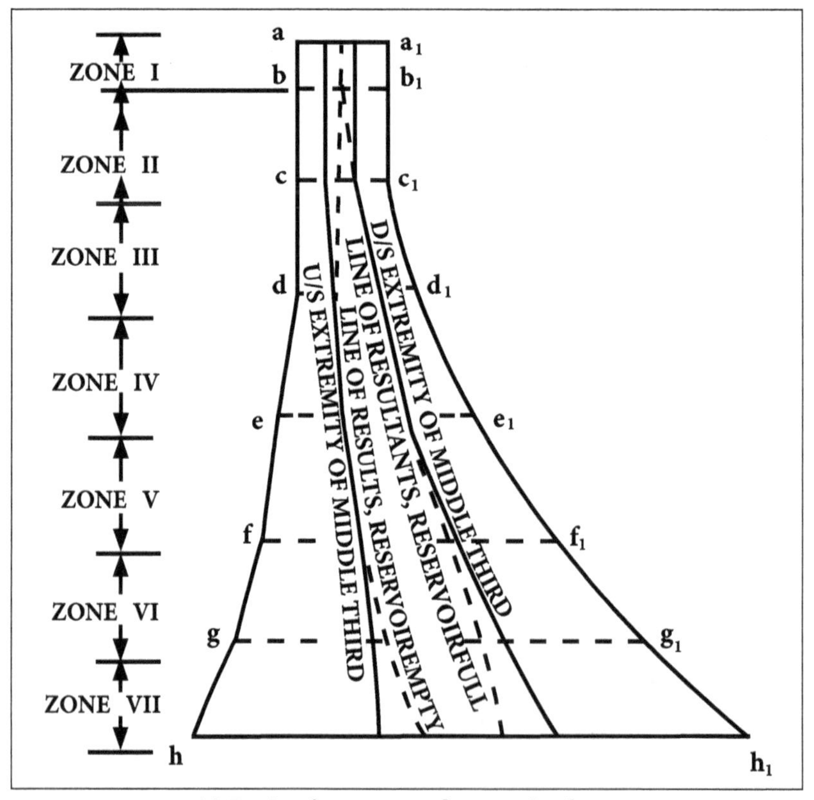

(1) Zoning for non-overflow gravity dam.

Zone I: This is the portion aa_1b_1b of the dam situated above the maximum reservoir level (or bottom of ice sheet where ice sheet exists). When no ice exists, the height of zone I is controlled by free board requirements and the width is determined by practical considerations or economy for the whole section. In the case of ice sheet the height of block I is fixed on the consideration of sliding of the zone due to ice pressure.

Zone II: This is the portion of the dam in which both the u/s and d/s faces remain vertical. The position of the bottom plane cc_1 of zone II is such that the resultant forces, when the reservoir is full, passes through the outer third point of the plane cc_1. However, when reservoir is empty, the line of resultant forces lies well within the middle third, throughout the height of zone II.

Zone III: In this Zone, the u/s face continues to be vertical while the d/s face is inclined. The line of resultant continues to coincide with the d/s extremity of middle third when the reservoir is full. The position of bottom plane dd_1 of zone III is located such that when the reservoir is empty the line of resultant passes through the u/s middle third point.

Zone IV: In this zone, the upstream face also begins to batter so that the lines of resultants lie along the corresponding extremities of middle third. The position of lower limit plane ee_1 of this zone is governed by the criterion that the maximum inclined pressure at d/s toe, for reservoir full condition is just equal to the allowable limit.

The design of zone IV, specially the height as well as both u/s and d/s slopes, are determined by trial, by dividing zone IV into number of convenient blocks, till the bottom of zone IV is reached. It should be noted that low dams lie within the limits of zone IV. Zones V, VI and VII are applicable only for high dams.

Zone V: In this zone, the d/s slope is flattened so that the maximum inclined pressure at d/s toe, under reservoir full condition remains within the working stress. Thus the resultant for reservoir full condition remains well within middle third. For the reservoir empty condition, the resultant continues to intersect the u/s extremity of the kern. The lower limit of zone V is marked by plane ff_1, where the inclined compressive stress on the u/s heel reaches the permissible limit, for the reservoir empty condition.

Zone VI: In this zone, the conditions of the design are determined by the maximum pressures at both u/s and d/s faces under reservoir empty and reservoir full conditions, respectively. The line of resultants under both the conditions lie well within the middle third. The position of the bottom plane gg1 is reached when the inclined pressure at d/s toe just reaches its maximum value.

Zone VII: This is the zone situated below zone VI, in which maximum compression at the downstream toe exceeds the working limit. This zone is usually eliminated by the revision of the entire design.

That is, if the height of dam is so large that it is more than the position of plane gg1 of zone VI, various changes are made in the upper zones so that the height of the dam lies within the zone VI. If this is not possible, then the height of the dam is reduced or superior material is utilized so that the height is accommodated within VIth zone.

Single Step Method of Design of High Dams

For high dams, going beyond zone IV, it is found that the shape of u/s and d/s slopes are sometimes obtained of unusual shape, Figure (2a) shows a typical section of high dam obtained by multiple step method. The u/s face has steep slope while the d/s slope has convex shape outwards. A convexed face under compression, whether smoothly curved or polygonal, may be subjected to tensile stresses on surface parallel to the face. Such shape is usually not desirable since the outer layer of such a section tends to buckle out-ward.

Hence, the whole section may have to be revised in such a way that such a reversed curvature is avoided. The simple method would be to start the redesign work from the top, converting the whole dam into a single block controlled by the rules of zone VI.

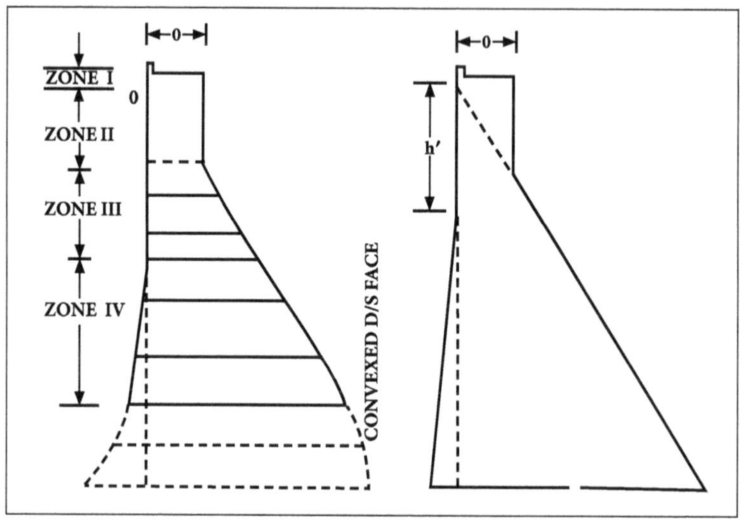

(a) Multiple step method (b) Single step method
(2) Single step method of Design of High Dams.

Figure (2b), shows the section of the same high dam designed by the single step method. In this method, the u/s slope is kept vertical for some depth, to be determined by trial. It may preliminarily be fixed by Equation

$$h' = 2a\sqrt{\rho} - c$$

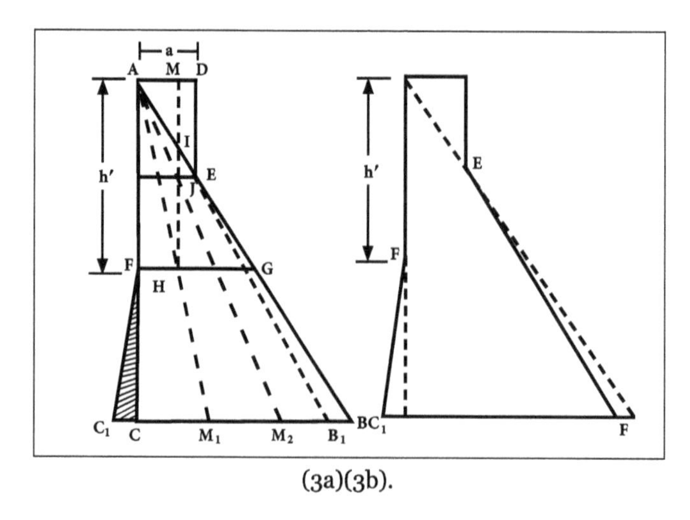

(3a)(3b).

The plane FG [Figure (3a)] passes through the intersection of the middle third line AH and the centroidal line MH of the top width triangle. After this height, both the u/s and d/s faces are given such slopes that the maximum inclined pressure at both u/s and d/s face under the conditions of reservoir empty and reservoir full conditions, reach their maximum values simultaneously.

This is to be accomplished by trial and error. When this is accomplished, computations will show that all the stability requirements for the reservoir full or empty conditions are satisfied at all points above the base.

Figure (4) shows the sections of a high dam, designed both by the single step method and multiple step method. It is quite clear from the two sections that the multiple-step design is more economical for the upper portion of the dam. The single-step-method section is under stressed at all the points except at the base. Hence if the height of the dam is less, multiple-step-design method would give substantial saving in material.

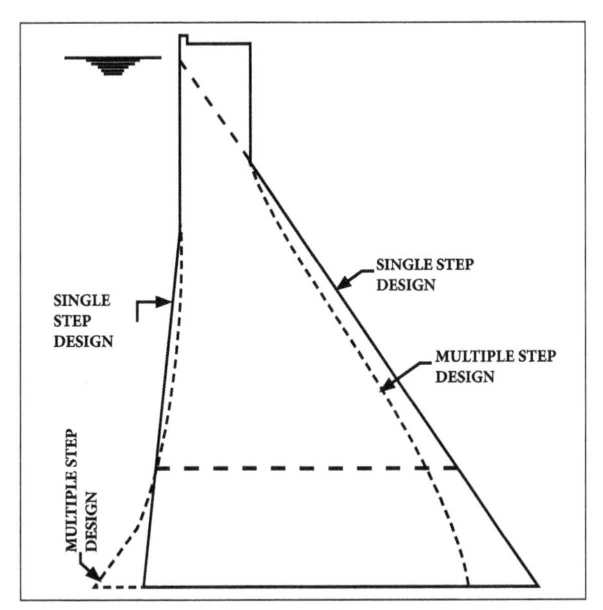

(4) Comparison of Multiple-Step method and Single-Step method for the Design of High Dams.

We can write the following conclusions:

- Dams of lesser heights can be designed economically only by multiple step-design method.
- It may be economical to increase the masonry strength through the use of more expensive materials, thus keeping out of zones V and VI.
- High dams beyond zones VI are designed by single step method so that convex-curvature of Ws face is avoided.

Alternative Method for the Design of High Gravity Dam

We have seen earlier that if the height of the dam is greater than that given by equation $h = \frac{f}{w(p+1)}$, the dam is known as a high dam. The design of such a dam, below the limiting height, is done by dividing the height into a number of suitable strips as shown in Figure (5).

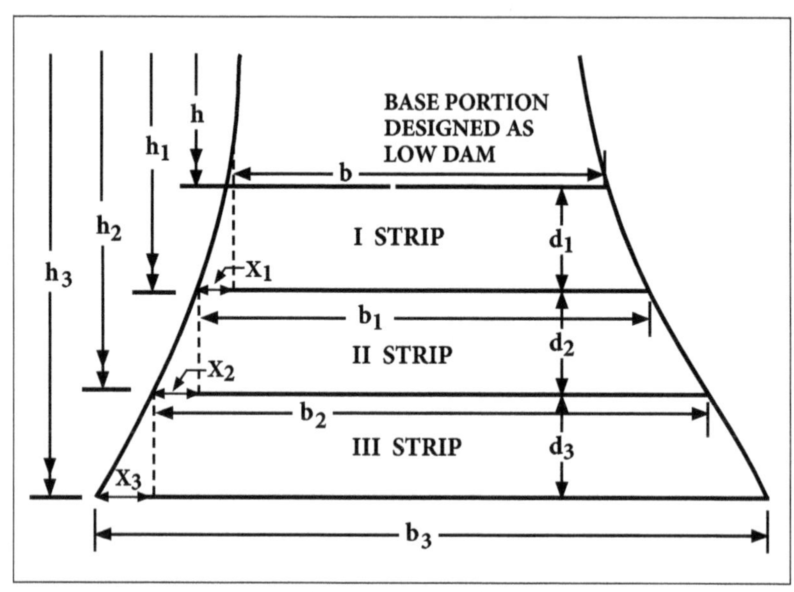

(5) Design of High Gravity Dam.

Design of First Strip

Let,

b_1 = Base width of the first strip.

b = Base width of the low dam.

h_1 = Height of the dam portion from H.F.L. to the bottom of the first strip.

W_1 = Total vertical load of dam and water, above the base of the first strip.

d_1 = Height of the first strip.

W = Total weight of dam portion and water, on the top of the first strip.

Then the base width b_1, sufficient to keep the maximum compressive stress within the safe value f given by,

$$b_1 = \sqrt{\frac{wh_1^3}{f}\left(1+\frac{w^2h_1^4}{4W_1^2}\right)} \qquad ...(1)$$

The projection x_1 on the u/s of the heel is given by the following expression:

$$\frac{pwd_1}{24}\left\{3b^2 - b_1^2 + 6x_1(b+b_1) + 2bb_1\right\} - \frac{wx_1}{12}(h+h_1)(2b_1 - 3x_1)$$

$$-w\left(\frac{b_1-b}{3}-x_1\right)=0 \qquad \ldots(2)$$

Design for Second Strip

Using suffix 2 for the second strip, the corresponding expressions for b_2 and x_2 are as follows,

$$b_2=\sqrt{\frac{wh_2^3}{f}\left(1+\frac{w^2h_2^4}{4W_{2.}^2}\right)} \qquad \ldots(3)$$

And,

$$\frac{pwd_1}{24}\left\{3b_1^2-b_2^2+6x_2\left(b_1+b_2\right)+2b_1b_2\right\}-\frac{wx_2}{12}\left(h_1+h_2\right)$$

$$\left(2b_2-3x_2-w_1\left(\frac{b_2-b_1}{3}-x_2\right)\right)=0$$

Similarly, the remaining strips are designed, till the base of the dam is reached.

Typical Section of Low Gravity Dam

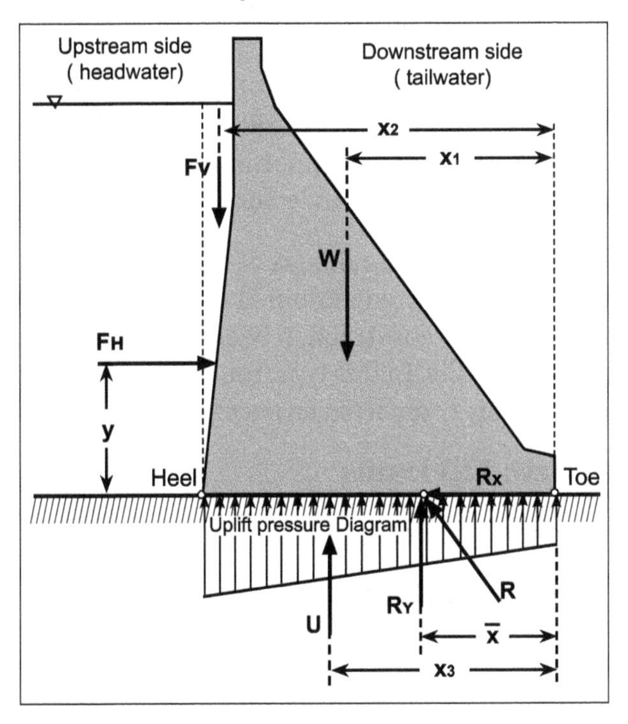

3.5 Earth Dams: Types and Causes of Failure

Earth Dams

Earth dams for the storage of water for the purpose of irrigation have been built since the earliest times. These dams were limited in height but not necessarily in extent. Now Earth dams are being built to unprecedented heights. Sites which have been considered unfit earlier for the construction of darns are now being exploited.

The Development of soil mechanics, i.e., the study of behavior of earth dams and the development of better construction techniques have been helpful in creating confidence to build higher dams with improved designs and more details. The result is that, today the highest dam in the world is an earth dam. The highest earth/rock fill dams in the world are Roguni U.S.S.R (335 m) Nurek, U.S.S.R. (300 m), Mica, Tehri India (260 m), Canada (244 m) and Oroville, U.S.A. (235 m).

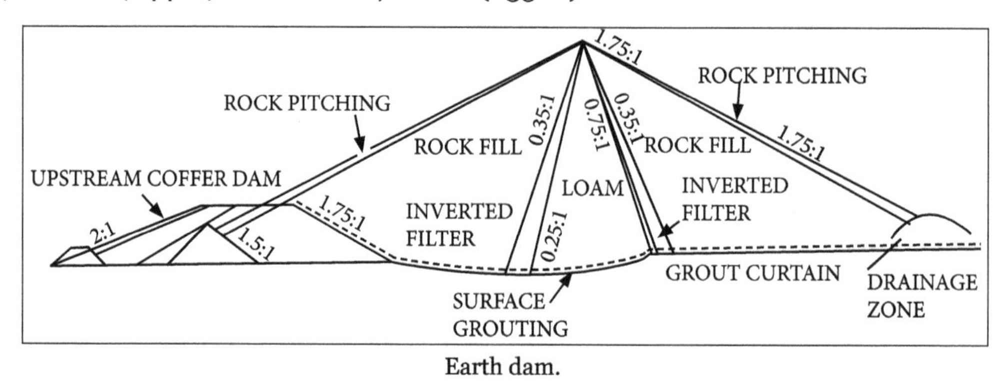

Earth dam.

In spite of these developments it is very difficult to establish mathematical solutions to the problems of design and components are still guided by experience and judgment. For a realistic design of an earth dam it is important that the foundation conditions and materials of construction are thoroughly investigated.

It is also necessary that the controlled methods of construction are used to achieve necessary degree of compaction at the predetermined moisture. This type of construction is being used almost entirely for the construction of earth dams to the exclusion of the hydraulic and the semi-hydraulic fills. In this type, the major portion of the embankment is built in successive mechanically compacted layers of 150 mm to 220 mm thickness each.

Classification of Earth Fill Dams

Earth fill dams are classified by number of factors:

1. Based on the Method of Construction

- Rolled fill earth dams.
- Hydraulic fill dam.

2. Based on Mechanical Characteristics of Earth Materials Making the Section of the Dam

- Homogeneous earth dams.
- Non-Homogeneous earth dams.
    - Non-homogeneous with inclined and impervious zone of artificial material.
    - With impervious zone of soil and also with low permeability.
    - With central core soil material of low permeability.
    - With a central thin diaphragm of impervious material.

Rolled Fill Earth Dams

In this type of dams, the successive layers of the moistened or damp soils are laid one over the other. Each layer that are not exceeding 20 cm in thickness is properly consolidated at optimum moisture content, only then the next layer laid.

Hydraulic Fill Dams

In this type of dams, the construction, the excavation, the transportation of the earth is done by the hydraulic methods. Outer edges of the embankments are kept slightly higher than the middle portion of the each layer. During construction, a mixture of the excavated materials in the slurry condition is pumped and discharged at the edges.

This slurry of the excavated materials and water consists of coarse and fine materials. When it is discharged near the outer edges, the coarser materials settle first at the edges, while the finer materials move to the middle and then settle there. Fine particles are deposited in the central portion to form a water tight central core. In this method, compaction is not required.

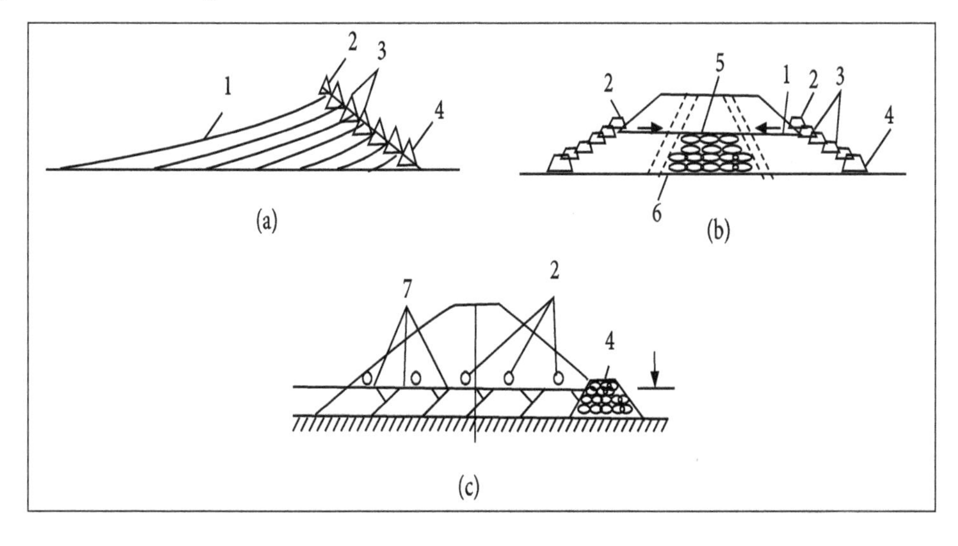

Homogeneous Earthen Dam

These dams are constructed with the uniform and homogeneous materials. It is suitable for low height dams. These dams are generally constructed with soil and grit mixed in the proper ratios. The seepage action of such dams are unfavorable, therefore for safety in case of rapid drawdown, the upstream slope is kept relatively flat.

Homogeneous section is modified by constructing the rock toe at the downstream lower end and providing horizontal filter drain.

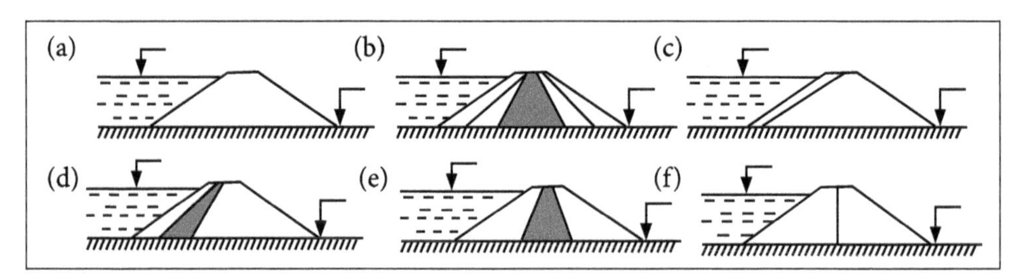

Zoned Earth Dams

These are the dams with the central portions termed as core or hearting made from the materials which are relatively impervious. The thickness of the core wall is made sufficiently thick to prevent it from leakage of water through the body of dam.

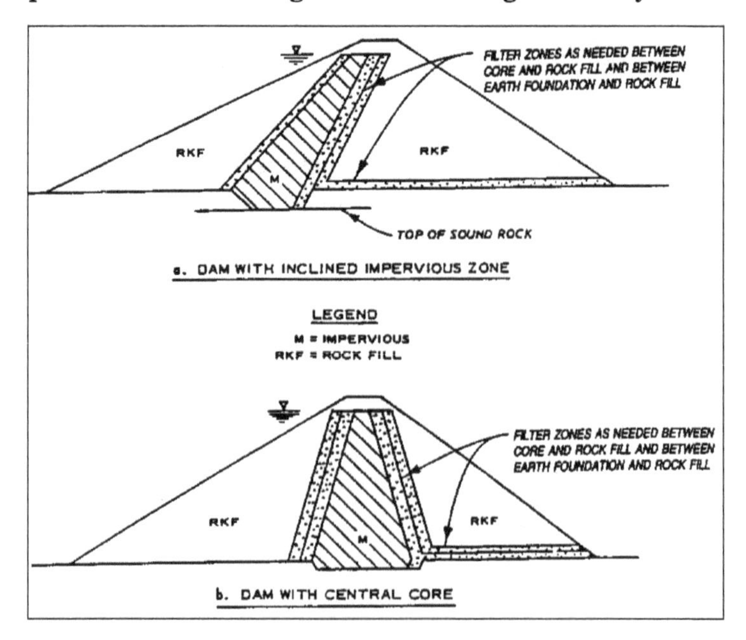

Dam with a Diaphragm

This type of dam is constructed with a thin impervious diaphragm in the central part to prevent it from seepage of water. The thin impervious diaphragm can be made by means of impervious clayey soil, cement concrete or masonry or any of the impervious material. The diaphragm may be constructed in the central portion or on the upstream face of the dam. The main difference in zoned and diaphragm type of dams depend on the thickness of the impervious core or diaphragm. The thickness of the diaphragm is not more than 10 m.

Design Criteria of the Earth Dams

Based on experience of the failures, following main design criteria can be laid down for the safety of an earth dam.

1. To prevent hydraulic failures the dam should be so designed that erosion of the embankment is always prevented.

This implies that the following conditions are strictly satisfied:

- Spillway capacity is sufficient to pass a peak flow.
- Overtopping by wave action at maximum water level is prevented.
- The original height of structure is sufficient to maintain the minimum safe freeboard after settlement has occurred.
- Erosion of embankment due to the wave action and the surface run-off does not occur.
- The crest should be wide enough to withstand wave action and earthquake shock.

2. To prevent the seepage failures, the flow of water through the body of the dam and its foundation must not be sufficiently large in quantity to defeat the purpose of the structure nor at a pressure sufficiently high to cause piping.

This implies that:

- Quantity of seepage water through the dam section and foundation should be limited.
- The seepage line should be well within the downstream face of the dam to prevent sloughing.
- Seepage water through the dam or foundation should not remove any particle or in other words cause piping. The driving force depends upon the pressure gradient while the resisting force depends upon the strength characteristics of the boundary material.
- There should not be any leakage of water from the upstream to downstream face. Such leakage may occur through conduits, at joints between earth and concrete sections or through the holes made by aquatic animals.

3. To prevent the structural failures, the embankment and its foundation should be stable under all conditions.

This implies that:

- The upstream and downstream slopes of embankment must be stable under all loading conditions to which they shall be subjected including the earthquake.
- The foundation shear stresses must be within the permissible limits of shear strength of the material.

Causes of Failures of Earth Dams

Like most other damages to engineering structures, earth dam failures are caused by improper design frequently based on insufficient investigations and lack of care in construction and maintenance.

Failures of earth dams may be grouped into the following basic causes:

- Hydraulic failures.
- Seepage failures.
- Structural failures.

1. Hydraulic Failures

They account for about one third of the failure of dams and are produced by surface erosion of the dam by water. They include wash-outs from overtopping wave erosion of upstream face, scour from the discharge of the spillway etc. and erosion from rainfall.

2. Seepage Failures

Seepage of water through the foundation or embankment has been responsible for more than one third of earth dam failures. Seepage is inevitable in all earth dams and ordinarily it does no harm. Uncontrolled seepage may however, cause erosion within the embankment or in the foundation which may lead to piping.

Piping is the progressive erosion which develops under the dam. It begins at a point of concentrated seepage where the gradients are sufficiently high to produce erosive velocities. If forces resisting erosion i.e. cohesion, inter-locking effect, weight of soil particles, action of downstream filter etc. are less than those which tend to cause, the soil particles are washed away causing piping failure.

Seepage failures are generally caused by:

- Pervious foundations.
- Leakage through embankments.

- Conduit leakage.
- Sloughing.

Pervious Foundations

Presence of strata and lenses of sand or gravel of high permeability or cavities and fissures in the foundation may permit concentrated flow of water from the reservoir causing piping. Presence of buried channels under the seat of dam have also been responsible for this.

Leakage Through Embankments

The following are the common causes of embankment leaks which lead to piping:

- Poor construction control which includes insufficient compaction adjacent to outlet conduits and poor bond between embankment and the foundation or between the successive layers of the embankment.
- Cracking in the embankment or in the conduits caused by foundation settlement.
- Animal burrows.
- Shrinkage and dry cracks.
- Presence of roots, pockets of gravel or boulders in the embankment.

Conduit Leakage

Conduits through the dam have been responsible for nearly one third of the seepage failure and more than one eighth of all failures. Failures are of two types:

- Contact seepage along the outside of the conduit which develops into piping.
- Seepage through leaks in the conduit which may also develop into piping.

Contact seepage along the conduit wall is caused either by a zone of poorly compacted soil or small gap between the conduit and remainder of the embankment. Seepage through poorly compacted zones soon develops into piping. Conduit cracking is caused by differential settlement or by overloading from embankment.

Sloughing

Failure due to sloughing takes place where downstream portion of the dam becomes saturated either due to choking of filter toe drain or due to the presence of highly pervious layer in the body of the dam.

The process begins when a small amount of material at the downstream toe is eroded and produces a small slide. It leaves a relatively steep face which becomes saturated by seepage from the reservoir and slumps again, forming a higher and more unstable face. This process is continued until the remaining portion of the dam is too thin to withstand the water pressure and complete failure occurs.

3. Structural Failures

Structural failures of the embankment or its foundation account for about one fifth of the total number of failures. Structural failures may result in slides in foundation or embankment due to various causes which are as follows,

Foundation Failures

Faults and seams of weathered rocks, shale, and soft clay strata are responsible for the foundation failure in which the top of the embankment cracks and subsides and the lower slope moves outward and large mud waves are formed beyond the toe. Another form of foundation failure occurs because of excessive pore water pressure in confined seams of silt or sand.

Pore water pressure in the confined cohesionless seams, artesian pressure in the abutments or consolidation of clays interbedded with the sands or silt, reduces the strength of the soil to the extent that it may not be able to resist the shear stresses induced by the embankment. The movement develops very rapidly without warning. Excess settlement of foundation may also cause cracking of the embankment.

Types of Failures-Earth Dams

Slides in Embankment

An embankment is subjected to shear stresses imposed by pool fluctuations, seepage or earthquake forces. Embankment slides may occur when the slopes are too steep for the shear strength of the embankment material to resist the stresses imposed. Usually the movement develops slowly and is preceded by cracks on the top or the slope near the top.

Failure of this type are usually due to faulty design and construction. In high dams slope failure may occur during dissipation of pore pressure just after construction. The upstream slope failure may occur due to sudden drawdown as shown in figure. The downstream slope is critical during steady seepage condition.

Section of Dam

Embankment dams are constructed of either earth fill or a combination of earth and rock fill. Therefore, embankment dams are generally built in areas where large amount of earth or rocks are available. They represent 75% of all dams in the world.

Gravity dams depend entirely on their own weight to resist the tremendous force of stored water. In the earlier times, some dams have been constructed with masonry blocks and concrete. Today, gravity dams are constructed by mass concrete or roller compacted concrete.

Arch dams are concrete dams that curve upstream toward the flow of water. They are generally built in narrow canyons, where the arch can transfer the water's force to the canyon wall. Arch dams require much less concrete than gravity dams of the same length, but they require a solid rock foundation to support their weight.

Buttress dams depend for support on a series of vertical supports called buttresses, which run along the downstream face.

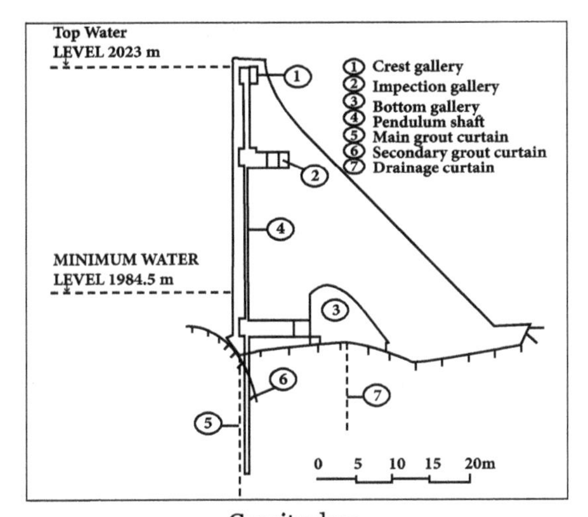

Gravity dam.

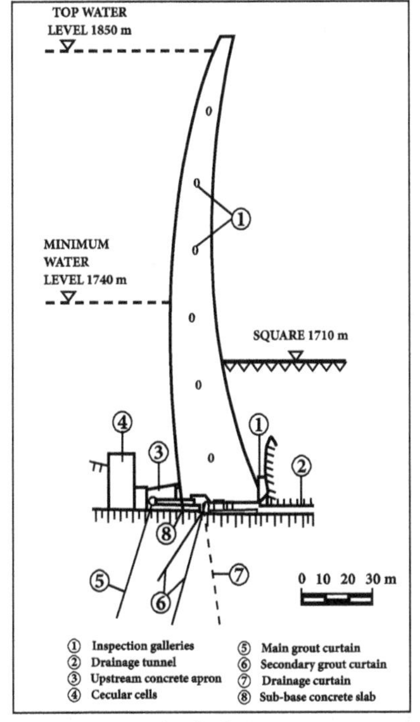

Arch dam.

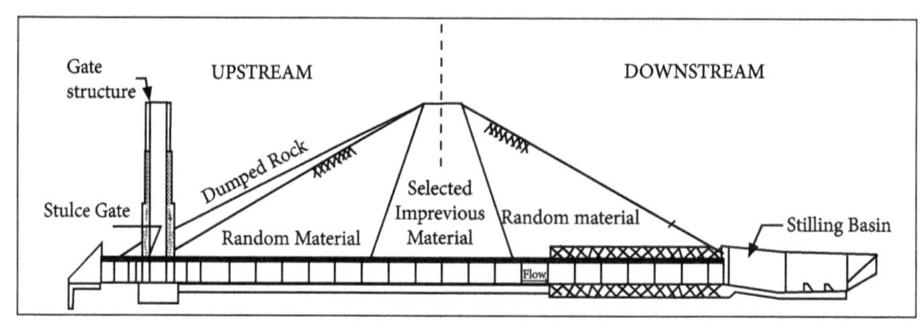

Embankment dam.

3.5.1 Preliminary Section of an Earth Dam, Seepage Control in Earth Dams

Preliminary Design Criteria

The preliminary design of earthen dam is done on the basis of past experiences. For designing purpose several parameters should be considered.

They are:

- Top Width.
- Free Board.
- Settlement Allowance.
- Casing or Outer Shell.
- Cut-off Trench.
- Downstream Drainage System.

1. Top Width: Minimum top width (W) should be such that it can enhance the practicability and protect it against the wave action and earth wave shocks. Sometimes it is also used for transportation purposes. It depends upon the height of the earthen dam and can be calculated as follows,

$$w = \frac{H}{5} + 3(\text{for very low dam})$$

$$w = 0.55\sqrt{H} + 0.2H(H \leq 30)$$

$$w = 1.65\sqrt[3]{H + 1.5}(H \leq 30)$$

Where,

H = the height of the dam (m), for Indian conditions it should not be less than 6 m.

2. Free board: It is the vertical distance between the top of the dam and the full supply level of the reservoir or the added height. It acts as a safety measure for the dam against high flow condition that is waves and runoff from storms greater than the design frequency from overtopping the embankment. The Recommended values of free board for different heights of earthen dams is given by U.S.B.R.

Recommended values of free board given by U.S.B.R. are:

If fetch length or exposure is given then the free board can also be calculated by Hawksley's formula:

$$h_w = 0.014 D_m^{0.5}$$

Where,

h_w = Wave height (m).

D_m = Fetch or exposure (m).

3. Settlement Allowance: It is the result of the settlement of the fill and foundation material resulting in the decrease of dam storage. It depends upon the type of fill material and the method and speed of construction. It varies from 10% of design height for hand compacted to 5% for machine compacted earth fill.

4. Casing or Outer Shell: Its main function is to provide stability and protection to the core. Depending upon the upstream and downstream slopes, a recommendation for the casing and outer shell slopes for different types of soils given by Terzaghi is presented in Table.

Table: Recommended Slopes of Earthen Dam:

Sl. No.	Types of material	u/s slope	d/s slope
1.	Homogeneous well graded material	$2\frac{1}{2}:1$	2:1
2.	Homogeneous coarse silt	3:1	$2\frac{1}{2}:1$
3.	Homogeneous silty clay or clay a) Height less than 15m b) Height more than 15m	$2\frac{1}{2}:1$ 3:1	2:1 $2\frac{1}{2}:1$
4.	Sand or and gravel with clay core	3:1	$2\frac{1}{2}:1$
5.	Sand or and gravel with R.C. Core well	$2\frac{1}{2}:1$	2:1

5. Cutoff Trench: It is provided to reduce the seepage through the foundation and also to reduce the piping in the dam. It should be aligned in a way that its central line should be within the upstream face of the impervious core. Its depth should be more than 1 m. Bottom width of cutoff trench (B) is calculated as:

$$B = h - d$$

Where,

h=Reservoir head above the ground surface (m).

d=Depth of cutoff trench below the ground surface (m).

6. Downstream Drainage System: It is performed by providing the filter material in the earthen dam which is more pervious than the rest of the fill material. It reduces the pore water pressure thus adding stability to the dam.

Three types of drains used for this purpose are:

- Toe Drains.
- Horizontal Blanket.
- Chimney Drains.

Seepage Control in Earth Dams

Regularly scheduled monitoring and inspection is essential to detect seepage and prevent dam failure. Inspections should be made periodically throughout the year. Frequency should be based on hazard classification of the dam.

Higher classified dams should be checked more common, compared to those that are lower hazard classified. At a minimum all dams should be visually inspected at least every six months, before a predicted major storm event, during or after severe rainstorms or snowmelts and inspected weekly after construction is complete and reservoir filling is ongoing and for at least two months after the reservoir has been filled.

Dam inspections performed on a regular basis are the most economical aid a dam owner can use to assure the safety and long life of the structure while reducing liability risks. If seepage is detected on a dam embankment or foundation, it should be closely monitored on a regular basis until it is corrected.

If seepage flow increases or embankment soils are showing signs of instability, corrective action should be taken quickly. Seepage problems at high hazard dams need to be corrected immediately. If the problems are neglected and permitted to progress, they may result in loss of life and property downstream of the dam.

A qualified geotechnical engineer or dam safety professional should be contacted for inspection and advice for all high hazard dam seepage problems. The type of controls deployed depends on the source, type and extent of seepage. Flow nets can also be installed as a method of studying the path that the seeping or moving water follows. Then the correct action can be taken.

If excessive water is flowing from soil piping or boils or if the water is carrying sediment, an engineer or safety professional should be contacted for inspection and recommendations for further action. The reservoir level should be lowered if serious piping or embankment sliding/sloughing is occurring and the cause of the condition corrected.

Sloughing and sliding due to seepage at the toe of the embankment may be corrected by removing the unstable soil and constructing a toe drain with filter out of permeable soil. Permeable soils and materials are commonly used at the toe of dams, to create drains to help control and prevent erosion. By using such materials, it allows the water to pass through without carrying the soil with it. The same thing can be done in locations where water is flowing but piping has not yet occurred.

Seepage, piping and boils in existing dams may be corrected or slowed, by intercepting the water before it exits on the downstream side of the dam. Some typical methods of intercepting include impermeable upstream blankets, cutoff trenches in the embankment, grout curtains, relief wells and toe drains. Impermeable upstream blankets or liners, are the most effective method, but require complete drawdown of the reservoir.

These blankets may consist of low-permeability soil or a synthetic geo-membrane. The blankets may also be deployed on the floor of the reservoir to prevent foundation seepage. All cracks and erosion rills on the embankment should be filled, re-graded and re-seeded. Borrowing rodents should be eliminated form dams and any damage created should be repaired by backfilling with a soil or filtered drain.

Safety Measures

The dam safety can be ensured if the following aspects are taken care of:

Hydraulic Failure

Such type of failure can be averted by providing:

- Adequate spillway capacity.
- Adequate freeboard so that dam safety is not endangered by overtopping during high floods.
- Proper maintenance of gates so that they are always operative and do not get clogged.

Seepage Failure

Seepage failure can be taken care by:

- Providing filter at the toe to minimize movement of the material.
- Seepage line is well within the body of the dam.
- Provision of settlement, after the composition of the dam, be made from a normal 1 % of height to a maximum of 6 %.

Structural Failure

In spite of best geological and foundation investigations done prior to dam construction, geological problems may arise such as induced seepage, earth tremors, slides, gougy seams and sloughing in the vicinity of dam and reservoir area surface during the construction or several years after the reservoir filling.

Periodic geotechnical inspection is essential for early detection and resolution of potential problems, besides provision of adequate rip-rap and its maintenance and draw-down within permissible limits.

Earthquake Failure

Necessary provisions shall be made in the design of a dam to account for additional forces due to earthquakes.

3.6 Spillways

The spillways are openings provided at the body of the dam to discharge the excess water or flood water safely when the water level rises above normal pool level.

Necessity of Spillways

- The height of dam is fixed according to the maximum reservoir capacity. The normal pool level is the one that indicates the maximum capacity of the reservoir. The water is never stored in reservoir above this level. The dam might fail by overturning, so for the safety of the dam the spillways are essential.
- The top of the dam is usually utilized by making road. The surplus water is not to be allowed till the dam reaches its top, so to stop the over topping by the surplus water, the spillways become extremely responsible.
- To protect the downstream base and floor of the dam from the effect of scouring and erosion, the spillways are provided in order to make the excess water flow smoothly.

Spillway Components

In general spillways comprise five distinct components namely:

- Discharge carrier.
- Entrance channel.
- Control structure.
- Outlet channel.
- Energy dissipator.

The entrance channel transfers water from reservoir to the control structure which regulates discharge from the reservoir. Water is then conveyed from reservoir to low-level energy dissipator on riverbed by the discharge conveyor. An energy dissipator is needed to reduce the high velocity of the flow to a non scouring magnitude.

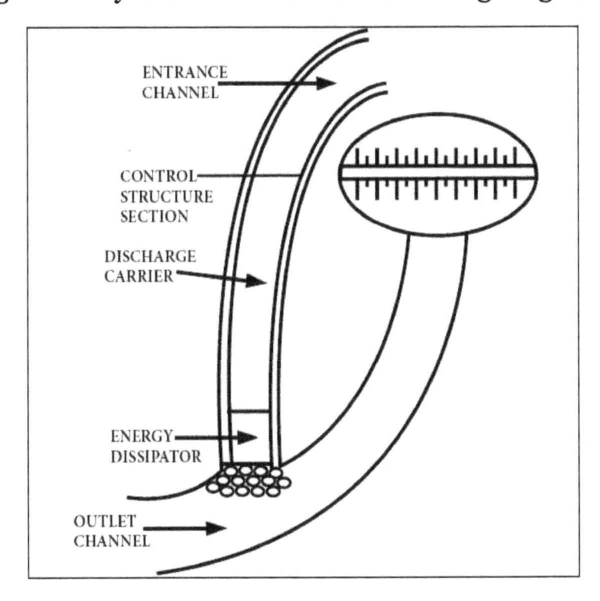

Factors Affecting Design

Safety considerations consistent with economy: Many failures of dams have resulted from the improperly designed spillway or spillways of inadequate capacity. The properly designed structure of the adequate capacity can be found to be only moderately higher in cost than a structure of inadequate capacity.

Hydrological and site conditions: The spillway design and its capacity depends on:

- Inflow discharge, its frequency and shape of the hydrograph.
- Height of the dam.

- Capacity curve.
- Geological and other site conditions.

Important topographical features, which affect spillways design are:

- Steepness of the terrain.
- Amount of the excavation and possibility of its use as the embankment material.
- The possibility of the scour.
- Stability of slopes, safe bearing capacity of the soils.
- Permeability of the soils.

For example, in case of narrow valley dams, side or chute channel spillway is very seldom possible, because of steepness of the banks and their insufficient stability.

Types of Spillway

Spillway is usually referred to as controlled or uncontrolled depending on whether spillway gates for controlling the flow have been provided or not. A free or uncontrolled spillway automatically releases water whenever the reservoir level rises above the overflow crest level.

The main advantage of an uncontrolled spillway is that it does not require constant attendance and operation of the regulating devices by an operator. Besides, there are no problems related to the maintenance and repair of the devices.

If it is not possible to provide a sufficiently long uncontrolled spillway crest or obtain a large enough surcharge head to meet the requirements of spillway capacity, one has to provide regulating gates. Such gates enable release of water, if required, even when the reservoir level is below the normal reservoir water surface level. Most common types of spillway are as follows:

- Free overfall (straight drop) spillway.
- Ogee (overflow) spillway.
- Side-channel spillway.
- Chute (or open channel or trough) spillway.
- Shaft (or morning glory) spillway.
- Siphon spillway.
- Cascade spillway.
- Tunnel (conduit) spillway.

1. Free Overfall Spillway

The flowing water drops freely from the crest of a free over fall spillway. At times, the crest is extended in the form of an overhanging lip to direct small discharges away from the downstream face of the overflow section.

The underside of the falling water jet is properly ventilated so that the jet does not pulsate. Such a spillway is better suited for a thin arch dam whose downstream face is nearly vertical. Since the flowing water usually drops into the stream bed, objectionable scour may occur in some cases and a deep plunge pool may be formed.

If erosion cannot be tolerated, plunge pool is created by constructing an auxiliary dam downstream of the main dam. Alternatively, a basin is excavated and is provided with a concrete apron. When tail-water depth is sufficient, a hydraulic jump forms when the water jet falls upon a flat apron. Free over fall spillways are restricted only to situations where the hydraulic drop from the reservoir level to tail-water level is less than about 6 m.

2. Ogee (Overflow) Spillway

An ogee spillway has a control weir whose profile is as shown in figure. The upper part of the spillway surface matches closely with the profile of the lower nappe of a ventilated sheet of water falling freely from a sharp-crested weir. The lower part of the spillway surface is tangential to the upper curve and supports the falling sheet of water.

The downstream end of the spillway is in the form of a reverse curve which turns the flow into the apron of a stilling basin or into the spillway discharge channel. An ogee spillway is generally used for concrete and masonry dams. It is ideally suited to wider valleys where sufficient crest length may be provided.

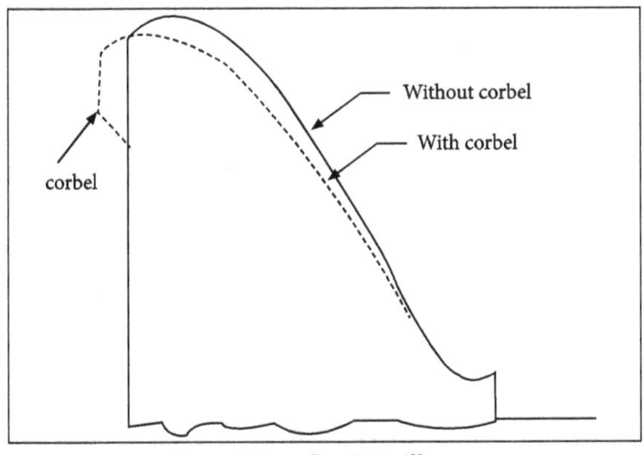

Ogee (Overflow) Spillway.

The profile of an ogee spillway is designed for a given design discharge (or corresponding design surcharge head or simply, design head). When the flowing discharge equals the design discharge, the flow adheres to the spillway surface with minimum interference from the boundary surface and air has no access to the underside of the water sheet.

The discharge efficiency is maximum under such condition and the pressure along the spillway surface is atmospheric. If the flowing discharge exceeds the design discharge, the water sheet tends to pull away from the spillway surface and thus produces sub-atmospheric pressure along the surface of the spillway.

While negative pressure may cause cavitation and other problems, it increases the effective head and increases the discharge. On the other hand, positive hydrostatic pressure will occur on the spillway surface, if the flowing discharge is less than the design discharge.

Model tests have indicated that the design head may be safely exceeded by about 50% beyond which cavitation may develop. Therefore, spillway profile may be designed for 75% of the peak head for the maximum design flood.

An upstream overhang, known as corbel, is added to the upstream face of the spillway. The effect of the corbel is to shift the nappe (and hence, the spillway profile) backward which results in saving of concrete. If the height of the vertical face of the corbel is kept more than 0.3 times the head over the crest, the discharge coefficient of the spillway will be practically the same as it would be if the vertical face of the corbel were to extend to the full height of the spillway.

3. Side-Channel Spillway

The control weir of a side-channel spillway is located alongside and approximately parallel to the upstream portion of the spillway discharge channel which itself may be either an open channel, a closed conduit or an inclined tunnel. The spillway discharge flows over the weir crest and falls into a narrow trough (i.e., upstream of the discharge channel) and takes an approximately 90°-turn before continuing into the spillway discharge channel.

The control structure, in plan, may be straight, curved, semi-circular or U-shaped. The overflow section may be broad-crested instead of ogee-shaped. One should note that when the flow in the side channel trough is subcritical, the incoming flow from the control structure (i.e., the overflow crest) will not cause high transverse velocities because of the low drop due to relatively higher depth of flow in the trough.

This would effect good diffusion and intermingling of the incoming flow with the trough water due to relatively low velocities of both the incoming flow and the flow in the trough. Therefore, there would be comparatively smooth flow in the side channel trough.

However, when the flow in the side channel trough is supercritical, the flow velocities in the trough would be high and the depth of flow is small, causing the incoming flow to have a relatively higher drop. Therefore, the intermixing of the high-energy transverse flow with the trough stream will be rough and turbulent producing violent wave action causing vibrations. Therefore, the flow in the side channel trough should be maintained at subcritical condition for good hydraulic performance.

Moreover, the amount of excavation would also increase for larger bed widths. While the flow in the side channel trough should preferably be subcritical, the flow in the discharge channel is supercritical and a control section downstream of the trough is provided by either constricting the channel width or raising the channel bottom.

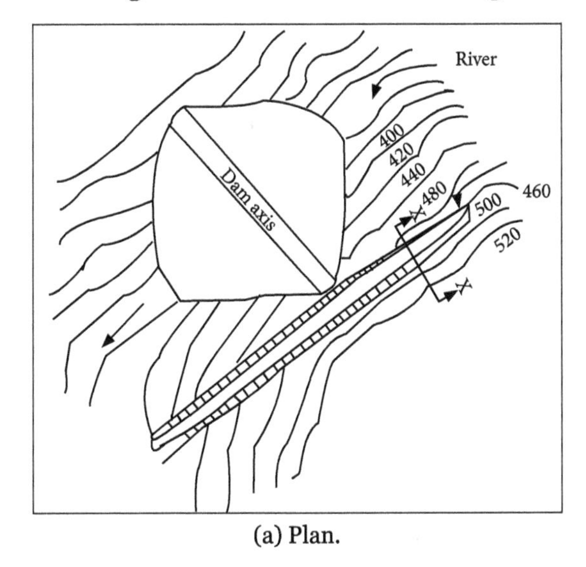

(a) Plan.

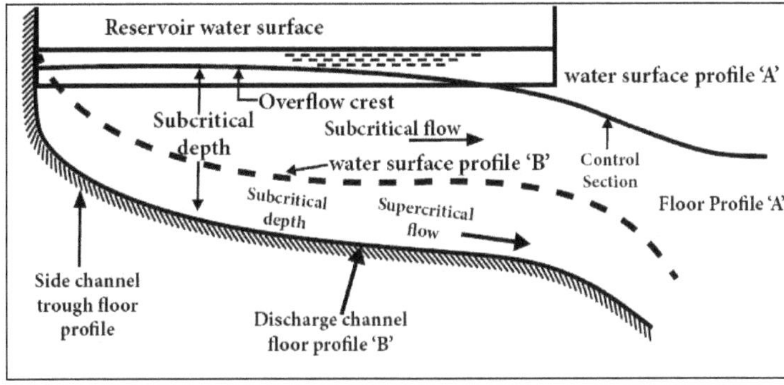

(b) Side channel profile.

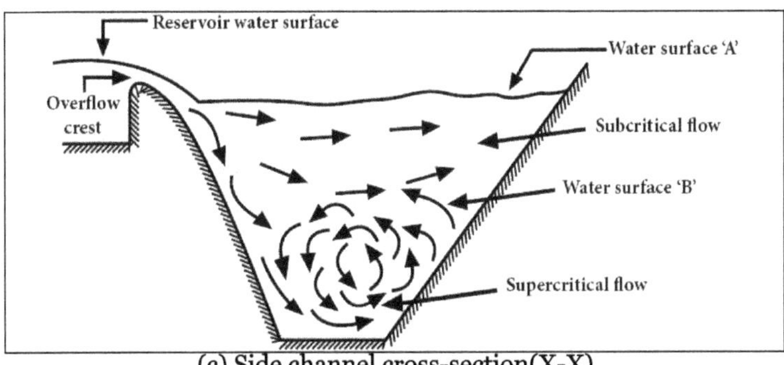

(c) Side channel cross-section(X-X)
Side-Channel Spillway.

4. Chute Spillway

In a chute (or trough) spillway, the spillway discharge flows in an open channel (named as 'chute' or 'trough') right from the reservoir to the downstream river. The open channel can be located either along the abutment of the dam or through a saddle. The channel bed should always be kept in excavation and its side slope must be designed to be stable with sufficient margin of safety.

As far as possible, bends in the channel should be avoided. If it becomes necessary to provide a bend, it should be gentle. The spillway control structure can be an overflow crest or a gated orifice or some other suitable control device. The control device is usually placed normal or nearly normal to the axis of the chute. The simplest form of a chute spillway is an open channel with a straight centre line and constant width. However, often the axis of either the entrance channel or the discharge channel is curved to suit the topography of the site.

The flow condition varies from subcritical upstream of the controlling crest to critical at the crest and supercritical in the discharge channel. The chute spillway is ideally suited with earth fill dams because of:

- Simplicity of their design and construction.
- Their adaptability to all types of foundation ranging from solid rock to soft clay.
- Overall economy usually obtained by the use of large amounts of spillway excavation for the construction of embankment.

The chute spillway is also suitable for concrete dams constructed in narrow valleys across a river whose bed is erodible for which the ogee spillway becomes unsuitable.

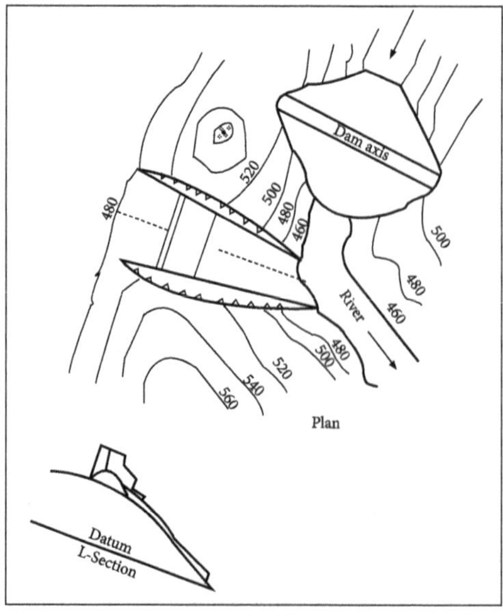

Chute Spillway.

5. Shaft Spillway

In a shaft spillway water enters a horizontal crest, drops through a vertical or sloping shaft and then flows to the downstream river channel through a horizontal or nearly horizontal conduit or tunnel. A rock outcrop projecting into the reservoir slightly upstream of the dam would be an ideal site for shaft spillway. Depending upon the level of the rock out crop and the required crest level, a spillway may have to be either constructed or excavated.

The diversion tunnels, if used for river diversion purposes during construction, can be utilized for discharge tunnels of the spillway. Radial piers provided on the spillway crest ensure radial flow towards the spillway and also provide support to the bridge which would connect the spillway with the dam or the surrounding hill.

A shaft spillway with a funnel-shaped inlet is called a "morning glory" or "glory hole" spillway. One of its distinguishing characteristics is that near maximum capacity of the spillway is attained at relatively low heads. Therefore, a shaft spillway is ideal when maximum spillway discharge is not likely to be exceeded.

Because of this feature, however, the spillway becomes unsuitable when a flow larger than the selected design flow occurs. This disadvantage can be got rid of by providing an auxiliary or emergency spillway and using the shaft spillway as service spillway.

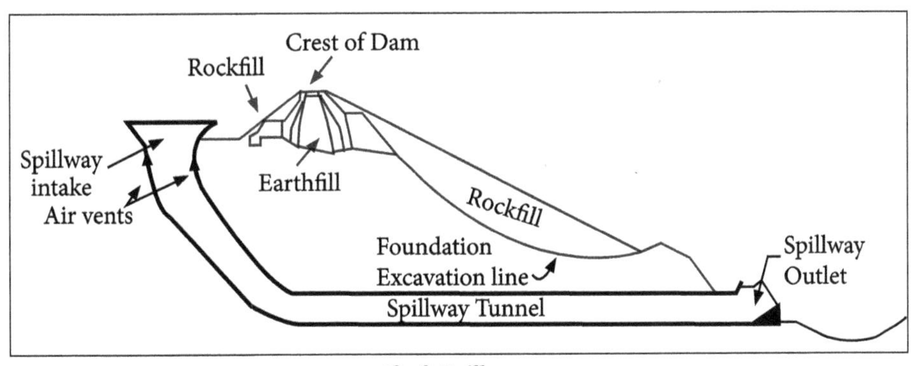

Shaft Spillway.

6. Siphon Spillway

A siphon spillway is essentially a closed conduit system which uses the principle of siphonic action. The conduit system is the shape of an inverted U of unequal legs with its inlet end at or below normal reservoir storage level. When the reservoir water level rises above the normal level, the initial flow of water is similar to the flow over a weir.

When the air in the bend has been exhausted, siphonic action starts and continuous flow is maintained until air enters the bend. The inlet end of the conduit is placed well below the normal reservoir water level to prevent ice and drift from entering the conduit. Therefore, once the siphonic action starts, the spillway continues to discharge even after the reservoir water level falls below the normal level.

As such, a siphon-breaking air vent is always provided so that siphonic action can be broken once the reservoir water level has been drawn down to the normal level in the reservoir. Siphon spillways can be either constructed of concrete or formed of steel pipe. The thickness of the wall of the siphon structure should, however, be sufficiently strong to withstand the negative pressures which develop in the siphon. Pressure at the throat section can be determined by the use of Bernoulli's equation.

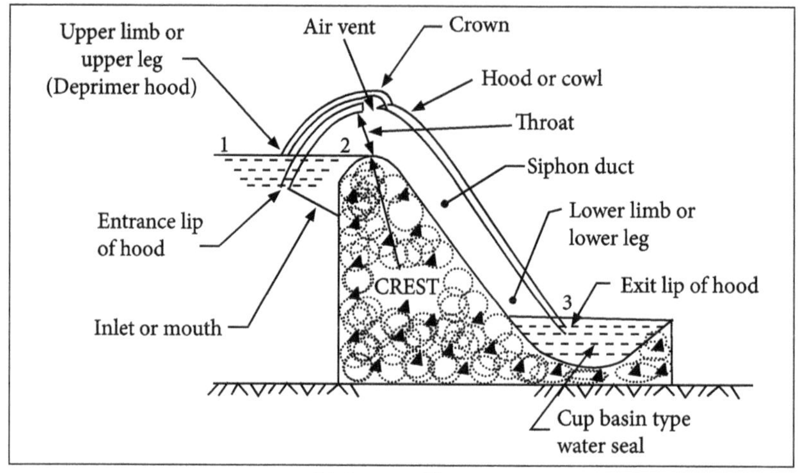

Siphon Spillway.

7. Cascade Spillway

In case of very high dams the kinetic energy at the toe of the dam will be very high and the tail water depth in the river may not be adequate for a single-fall hydraulic jump or roller bucket stilling basin. Narrow and curved canyons consisting of fractured rock would not be suitable for trajectory buckets. In such situations, especially for high earth and rock fill dams for which spillway is a major structure, possibility of providing a cascade of falls with a stilling basin at each fall as shown in figure below must be considered.

The cascade spillway (2, 8) is likely to be an ideal choice for a high rock fill dam for which the material has been obtained from a quarry located downstream of the dam so that the flood waters may be discharged over the quarry face. As the quarry would usually be excavated in benches, they may as well form the steps in the cascade.

A cascade spillway has been planned at the proposed 218 m high Tehri dam on the Bhagirathi river in the Ganga valley of the central Himalayas. At Darmouth dam in Australia, the spillway to pass 2700 m3/s of flood discharge is an unlined cascade in granite. The benches of the cascade will be 5 m high and of varying widths to suit the topography of the site.

Although, a cascade spillway would be an attraction during floods, it may not be always acceptable for environmental reasons which demand that the quarries be always located upstream of the dam and below the normal water surface level of the reservoir so as to cause minimum disfigurement of the land.

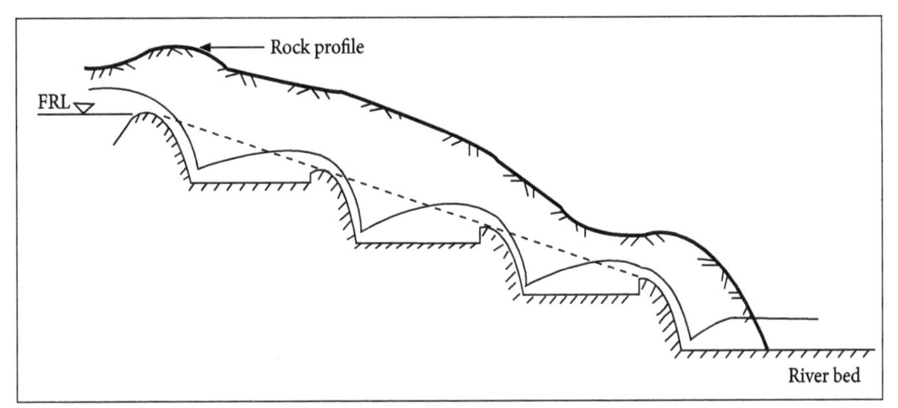

Cascade spillway.

8. Tunnel Spillway

A tunnel spillway discharges water through closed channels or tunnels laid around or under a dam. The closed channels can be in the form of a vertical or inclined shaft, a conduit constructed in an open cut and back-filled with earth materials or a horizontal tunnel through earth or rock. In narrow canyons with steep abutments as well as in wide valleys with abutments far away from the stream channel, tunnel spillways may prove to be advantageous. In such situations, the conduit of the spillway can be easily located under the dam near the stream bed.

Design Principles of Ogee Spillways

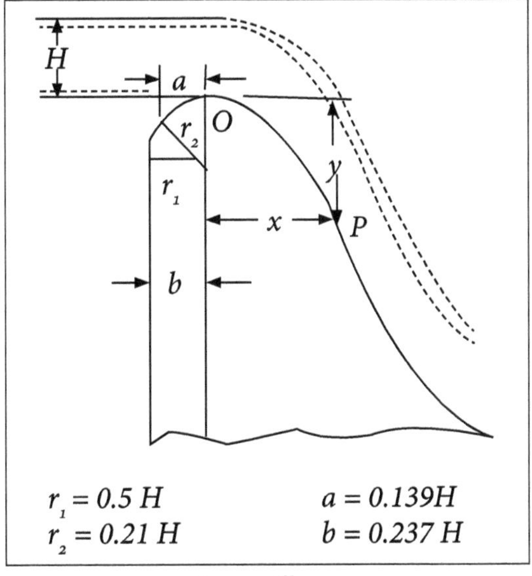

Ogee spillway.

The ogee spillway is the modified form of drop spillway. The downstream profile of the spillway is made to coincide with shape of the lower nappe of the free falling waterjet from the sharp crested weir. In this case, the shape of the lower nappe is similar to a projectile and hence downstream surface of the ogee spillway will have to follow the parabolic path where "o" is the origin of parabola.

The downstream face of the spillway forms concave curve from a point "T" and meets with the downstream floor. This point "T" is called as point of tangency. Therefore, the spillway takes the shape of the letter "S". Thus, this spillway is termed as ogee spillway.

Shape of the Ogee Spillway

The shape of the lower nappe is not the same for all the head of water above the crest of the weir. It differs with head of water. But for design of the ogee spillway the maximum head is considered. If the spillway runs with the maximum head, then the overflowing water just follows the curved profile of the spillway and there will be no gap between the water and the spillway surface and the discharge is maximum.

When the actual head becomes more than the designed head, the lower nappe does not follow ogee profile and gets separated from spillway surface. Hence, a negative pressure develops at the point of the separation. Due to the negative pressure, air bubbles are formed within the flowing water. These air bubbles are responsible for frictional force which causes much damage to spillway surface.

Again, if the head of water is less than designed head, the waterjet adheres to the body of the spillway and creases positive pressure which reduces the discharge through the spillway.

The shape of the ogee spillway has been developed by U.S Army Corps Engineers which is called as "Water-way experimental station spillway shape". The equation given by them is, $X_n = K \times H^{n-1} \times Y$, where, x and y are the coordinates of a point P on the ogee profile taking O as origin. K and n are the constants according to the slope of the upstream face of spillway.

Shape of u/s face of spillway	**K**	**n**
Vertical	2.0	1.85
1.3 (H:V)	1.936	1.836
1:1½ (H:V)	1.939	1.810
1:1 (H:V)	1.873	1.776

Therefore, for different values of y, the values of x are determined considering the slope of u/s face. The value of r_1, r_2, a and b are determined. The results are tabulated for the constructional guidance.

Problem

Let us design and sketch the shape of an Ogee type spillway for the following data using the empirical equation developed by US Army Corps Engineers.

Upstream Head, H = 20 m

Shape of the upstream face $= 1 : 1½(H : V)$

Values of K and n are 1.939 and 1.81 respectively.

X(m)	Y(m)
5	0.84
10	2.94
15	6.13
20	10.31
25	15.44

Solution

$$r_1 = 0.5H = 0.5 \times 20 = 10 \text{ m}$$

$$r_2 = 0.21H = 0.21 \times 20 = 4.24.5 \text{ m}$$

$$a = 0.139H = 0.139 \times 20 = 2.78\ 3 \text{ m}$$

$$b = 0.237H = 0.237 \times 20 = 4.74\ 5\text{m}$$

We know,

$$X^n = K \times H^{n-1} \times Y$$

$$X^{1.81} = Y \times 1.939 \times (20)^{1.81-1}$$

$$X^{1.81} = Y \times 21.95$$

$$Y = 0.159 \text{ m}$$

From the different values of X, the values of Y are found which is given above in a tabulated form: Finally the shape of Ogee type of spillway,

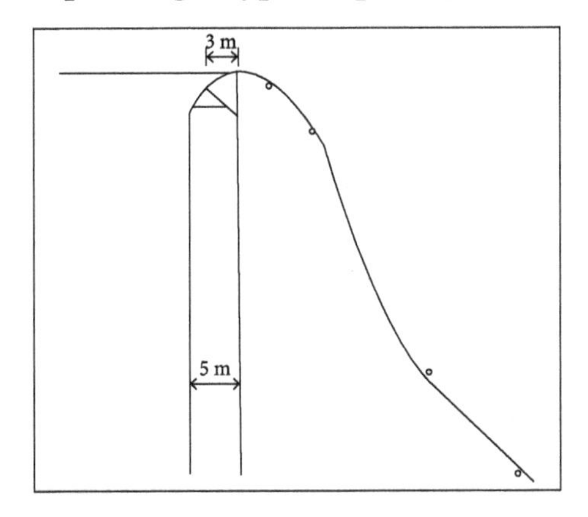

Free or uncontrolled overflow crest of a spillway is the simplest form of control as it automatically releases water whenever the reservoir water level rises above the crest

level. For such crests, there is no need of constant attendance and regulation of the control devices by an operator. Besides, the problems of maintenance and repair of the controlling device also do not arise. However, when sufficiently long, uncontrolled crest or large surcharge head for the required Spillway capacity cannot be obtained, a regulation gate may be necessary.

Such regulating devices enable the spillway to release storages even when the water level in the reservoir is below the normal reservoir water surface. Gates can be provided on all types of spillways except the siphon spillway. The installation of gates involves additional expenditure on initial cost and on their repair and maintenance. The selection of type and size of the controlling device depends on several factors, such as:

- Discharge characteristics of the device.
- Climate.
- Frequency and nature of floods.
- Winter storage requirements.
- The need for handling ice and debris.
- Special operating requirements such as presence of operator during periods of flood, the availability of electricity, operating mechanism and so on. In addition, economy, reliability, efficiency and adaptability of the regulating device must also the looked into.

Types of Spillways Crest Gates

The following types of regulating devices are generally used:

- Flashboards and stop logs.
- Rectangular lift gates.
- Radial gates.
- Drum gates.

These may be controlled either manually or automatically through mechanical or hydraulic operations.

Flashboards and Stop Logs

Flashboards and stop logs raise the reservoir storage level above a fixed spillway crest level when the spillway is not required to release flood. The flashboards usually consist of individual boards or panels of 1.0 to 1.25 m height. These are hinged at the bottom and are supported against water pressure by struts.

Stop logs are individual beams or girders set one upon the other to form a bulkhead supported in grooves at each end of the span. To increase the spillway capacity, the flashboards or stop logs are removed prior to the flood. Alternatively, they are designed and arranged so that they can be removed while being overtopped.

Flashboards and stop logs are simple and economical type of regulating devices which provide an unobstructed crest when removed. However, they have the following disadvantages:

- They present a hazard if not removed in time to pass floods, especially where the reservoir area is small and the stream is subject to flash floods.
- They require attendance of an operator or crew to remove them, unless designed to fall automatically.
- Ordinarily they cannot be restored to position while water flows over the crest.
- If they are designed to fail when the water reaches certain stage, their operation is uncertain and when they fail they release sudden and undesirably large outflows.
- If the spillway functions frequently, the repeated replacement of flash boards may be costly.

Vertical Lift Gates

These are usually rectangular in shape and made of steel which span horizontally between guide grooves in supporting piers and move vertically in their own plane. The gates are raised or lowered by an overhead hoist and water is released by undershot orifice flow for all gate openings.

Sliding gates offer large sliding friction due to water pressure and therefore, require a large hoisting capacity. The use of wheels (along each side of the gate) would reduce the amount of sliding friction and thereby permit the use of a smaller hoist. Vertical lift gates have been used for spans and heights of the order of 20 m and 15 m, respectively. At larger heights, however, the problem of a raised operating platform becomes important.

Radial (or Tainter) Gates

These are made of steel plates which form a segment of a cylinder which itself is attached to supporting bearing by radial arms. The cylindrical plate is kept concentric to the supporting pins so that the entire thrust of the water load passes through the pins and only a small amount of moment needs to be overcome in raising or lowering the gate.

The hoisting loads then include only the weight of the gate, the sliding friction and the frictional resistance at the pins. The small hoisting effort required for the operation of the radial gates makes hand operations at small installations possible. Besides, they require lesser head rooms than required by vertical lift gates. All these advantages make the radial gates more adaptable.

Drum Gates

Drum gates are hollow (and therefore, buoyant), triangular in section and made of steel plates. The drum gate is hinged at the upstream lip of a hydraulic chamber in the weir structure in which the gate floats. Water introduced into or drawn from the hydraulic chamber causes the gate to swing upwards or downwards. The inflow or outflow of water to the chamber is governed by controls located in the piers adjacent to the chambers.

Energy Dissipation Below Spillways: Stilling Basin

Different types of energy dissipators may be used along with a spillway, alone or in combination of more than one depending upon the energy to be dissipated and erosion control required downstream of a dam. Broadly the energy dissipators are classified under two categories – Stilling basins or Bucket Type. Each of these are further subcategorized as given below.

Stilling Basin Type Energy Dissipators

They may fundamentally be divided into two types:

1. Hydraulic Jump Type Stilling Basins

(i) Horizontal Apron Type

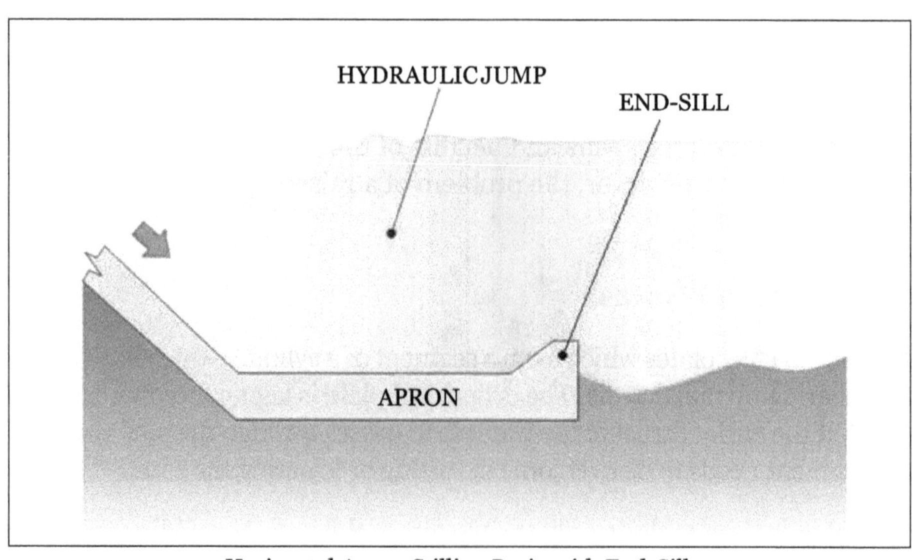

Horizontal Apron Stilling Basin with End-Sill

(ii) Sloping Apron Type

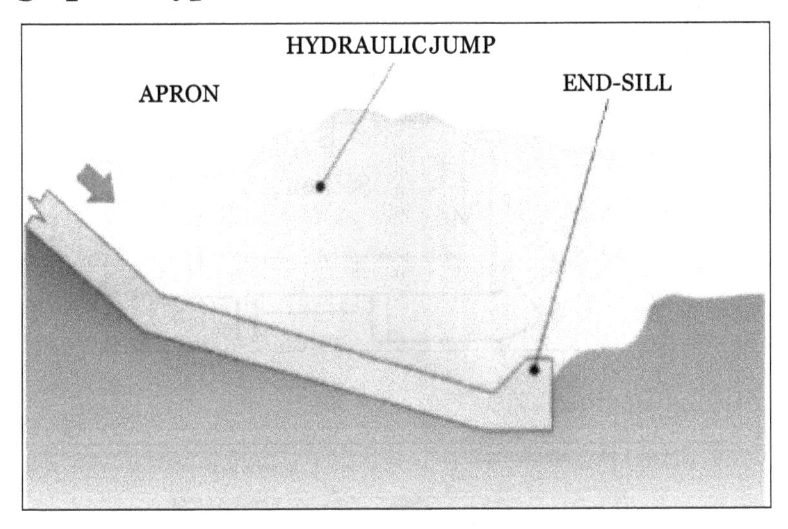

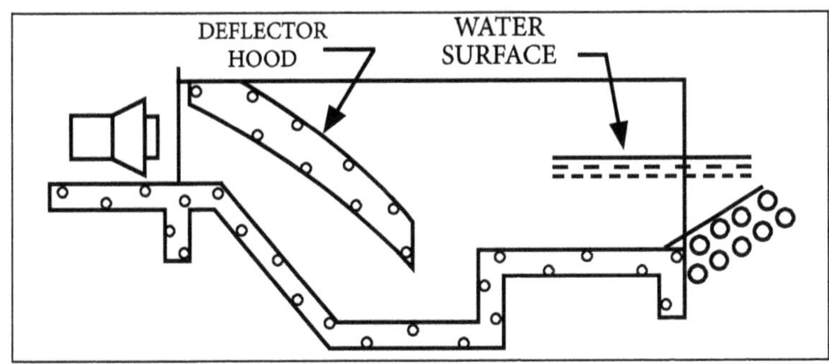

2. Jet Diffusion Type Stilling Basins

- Jet diffusion stilling basins.
- Interacting jet dissipators.

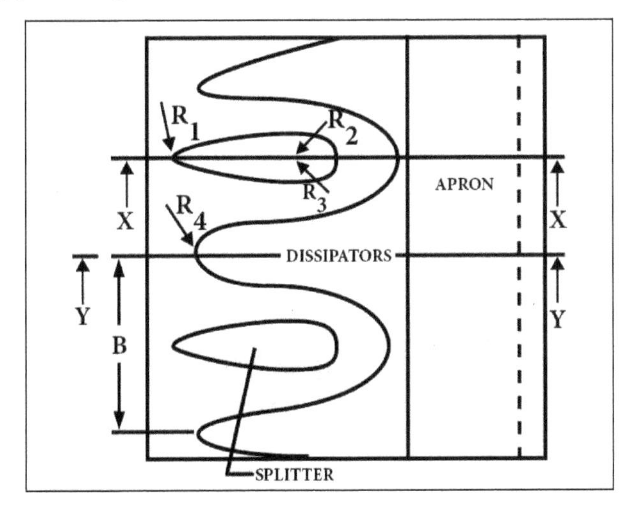

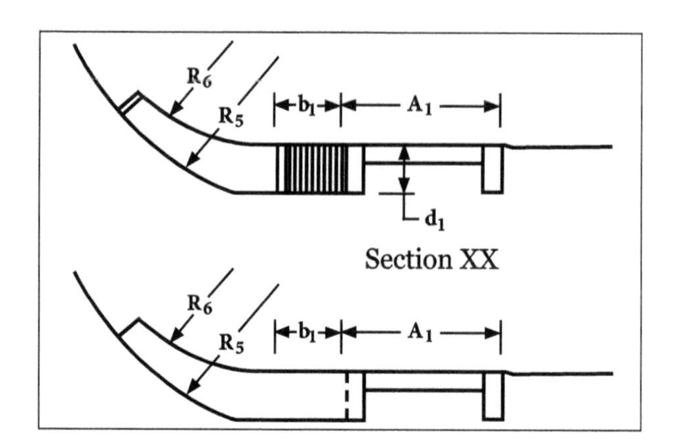

(iii) Free Jet Stilling Basins:

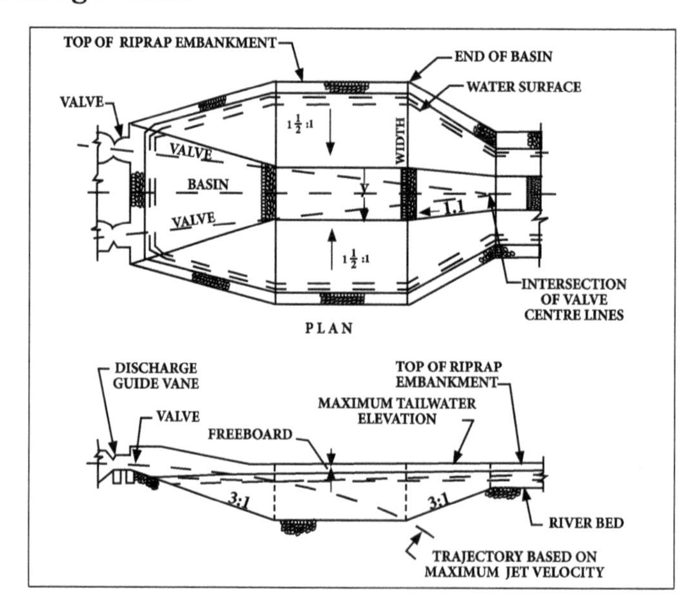

(iv) Hump Stilling Basins:

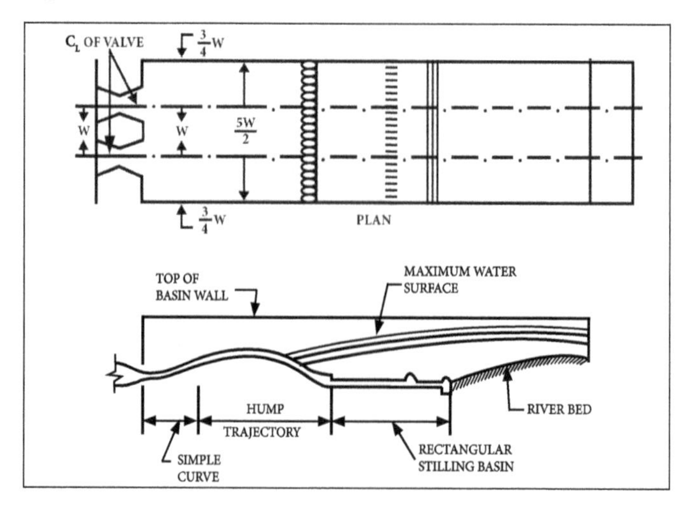

(v) Impact Stilling Basins:

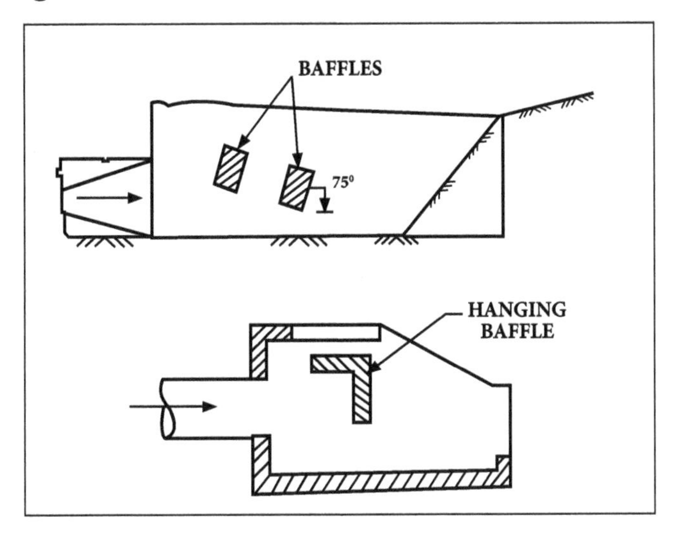

Bucket Type Energy Dissipators

This type of energy dissipators includes the following:

- Solid roller bucket.
- Slotted roller bucket.
- Ski jump (Flip/Trajectory) bucket.

Usually the hydraulic jump type stilling basins and the three types of bucket-type energy dissipators are commonly used in conjunction with spillways of major projects. Since energy dissipators are an integral part of a dam's spillway section, they have to be viewed in conjunction with the latter. Two typical examples have been shown in Figures though it must be remembered that any type of energy dissipator may go with any type of spillway, depending on the specific site conditions.

Permissions

All chapters in this book are published with permission under the Creative Commons Attribution Share Alike License or equivalent. Every chapter published in this book has been scrutinized by our experts. Their significance has been extensively debated. The topics covered herein carry significant information for a comprehensive understanding. They may even be implemented as practical applications or may be referred to as a beginning point for further studies.

We would like to thank the editorial team for lending their expertise to make the book truly unique. They have played a crucial role in the development of this book. Without their invaluable contributions this book wouldn't have been possible. They have made vital efforts to compile up to date information on the varied aspects of this subject to make this book a valuable addition to the collection of many professionals and students.

This book was conceptualized with the vision of imparting up-to-date and integrated information in this field. To ensure the same, a matchless editorial board was set up. Every individual on the board went through rigorous rounds of assessment to prove their worth. After which they invested a large part of their time researching and compiling the most relevant data for our readers.

The editorial board has been involved in producing this book since its inception. They have spent rigorous hours researching and exploring the diverse topics which have resulted in the successful publishing of this book. They have passed on their knowledge of decades through this book. To expedite this challenging task, the publisher supported the team at every step. A small team of assistant editors was also appointed to further simplify the editing procedure and attain best results for the readers.

Apart from the editorial board, the designing team has also invested a significant amount of their time in understanding the subject and creating the most relevant covers. They scrutinized every image to scout for the most suitable representation of the subject and create an appropriate cover for the book.

The publishing team has been an ardent support to the editorial, designing and production team. Their endless efforts to recruit the best for this project, has resulted in the accomplishment of this book. They are a veteran in the field of academics and their pool of knowledge is as vast as their experience in printing. Their expertise and guidance has proved useful at every step. Their uncompromising quality standards have made this book an exceptional effort. Their encouragement from time to time has been an inspiration for everyone.

The publisher and the editorial board hope that this book will prove to be a valuable piece of knowledge for students, practitioners and scholars across the globe.

Index

Printed in the USA
CPSIA information can be obtained
at www.ICGtesting.com
JSHW050858280224
58090JS00024B/139

9 781647 403560